THE KABUL KILL ZONE

THE KABUL KILL ZONE

An Air Force Officer's Account of Leadership, Loss, and Survival

John Spargur

Published by Ace Guardian Press

United States of America

ISBN: 979-8-9944372-0-9 (hardcover)

ISBN: 979-8-9944372-1-6 (paperback)

ISBN: 979-8-9944372-2-3 (eBook)

First edition

This book is a work of nonfiction. Names, identifying details, and certain events have been changed to protect privacy and operational security. The views expressed are those of the author and do not necessarily reflect the official policy or position of the United States Air Force, the Department of Defense, or the United States Government.

Printed in the United States of America

Dedication

To my family—whose legacy of service became the compass that guided my own.

To my grandfather **Clifford Wilson**, whose courage as a World War II Army infantryman carried him across Europe and into the heart of darkness to liberate prisoner-of-war camps. Your unwavering sense of duty showed me what real heroism looks like.

To my grandfather **John Spargur**, who stepped forward at seventeen to serve in the Navy during World War II. Your resolve taught me that service begins with a choice—one you made boldly, without hesitation.

To my father, **Dennis**, who manned the decks of the *USS Enterprise* during Vietnam. Your stories, sacrifices, and steady example shaped my understanding of honor long before I ever put on a uniform.

To my mother, **Judith**, who gave me the gift of language—who nurtured my love of reading and writing, and whose boundless support sustained me through every deployment. Your devotion and care for Jordan when I was away made my service possible.

To my wife, **Sandra**, my anchor and my home. Thank you for your steadfast love and support through the final years of my service.

And to my son, **Jordan**—you mean the world to me. Everything I have done, and everything I have endured, is illuminated by the hope I see in you. In my hardest moments downrange, when the days felt longest and the weight of war pressed hardest, it was the thought of you—your future, your smile—that steadied me. You were the light I carried with me, and the reason I kept going.

This book is for all of you. Your footsteps paved my path; your love kept me standing on it.

Contents

Author's Note

This memoir is my recollection—told as accurately and honestly as memory allows—of my time at Headquarters, International Security Assistance Force (HQ ISAF), in Kabul, Afghanistan, during my 2013 deployment. The experiences described here are my own, shaped by the moments I lived through and the decisions required in a place where survival was often measured in seconds.

This is not a story about heroism. It is a story about survival—immediate, unforgiving, and real. In Afghanistan, survival was not a metaphor for resilience; it was a calculation made each time I stepped beyond a blast-proof door, drove through a choke point, or listened for the first crack of incoming fire.

This book has no ulterior motive and no hidden agenda. It was not written to provoke controversy or assign blame. It is an honest account of the pressures faced by U.S. and coalition forces in a battlespace I came to know as the Kabul Kill Zone—a place where the front line was everywhere and nowhere at once. What follows reflects how I processed the chaos around me, the constraints of the system, and the realities of a war entering its final phase.

During my deployment, I served as the NATO Chief of Strategic Basing at HQ ISAF, working at the operational nerve center of America's longest war under the four-star general commanding coalition operations across Afghanistan. From that vantage point—part war room, part bureaucratic maze—I witnessed the deliberate beginning of the coalition's withdrawal.

There was, however, another dimension to the mission, one that existed beyond official briefings and formal chains of command. As a liaison coordinating with CIA operatives, I became a small node within an intelligence network tracking high-value targets across the country. Every base closure I planned, every adjustment to a transfer timeline, had to be weighed against ongoing intelligence operations. Some of the installations slated for closure supported surveillance and collection programs monitoring Taliban facilitators, IED networks, and insurgents operating inside the very institutions we were attempting to strengthen.

To protect the privacy, safety, and reputations of those referenced in this memoir, I have taken deliberate steps to safeguard identities. With limited exception, I have used only first names or nicknames, altered or omitted surnames, and changed call signs and other identifying details. In certain instances, timelines, characteristics, or events have been adjusted or combined for clarity or security, while preserving the substance and truth of the experiences described. Any names appearing unaltered are already a matter of public record.

This is my story—nothing more, nothing less. A record of uncertainty, pressure, and the split-second decisions that defined daily life in Kabul. I offer it with respect for all who served in Afghanistan, in the hope that it sheds light on a war most often fought beyond public view, moment by moment, breath by breath.

John Spargur
Kabul, Afghanistan

Prologue: Dangerous Shadows

The three men circled our vehicle like wolves scenting blood, deliberate and unafraid.

Through the bulletproof glass, I watched the one with the scarred face stop at my window, his eyes cataloging everything inside with predatory precision. The M4 assault rifles. The handguns. Our body armor. He wasn't just looking—he was calculating. This was a man who'd done this before.

"Sir, this doesn't look good," I said, my voice tighter than I wanted it to be. "Let's try a different route. We should leave. Now."

"No. We're going through this gate," General Washington replied, his jaw set with that stubborn determination I'd seen too many times before.

Twenty-five yards ahead, the makeshift checkpoint sat wrong in the deepening twilight. No sandbags. No proper barriers. Just an Afghan man with an AK-47, a black turban, and a metal arm gate stretched across a road I'd never been on before—a road we had no business being on as darkness swallowed Kabul whole.

The man had made a phone call the moment he spotted us. Then he'd simply waited, staring into the dark field beyond the wall. Not at us. Not acknowledging us. Just waiting.

That's when the other two appeared.

They'd materialized from the shadows with AK-47s, huddling with the first man in urgent conversation before all three began their slow circle around our heavily armored Toyota Land Cruiser. Their movements were too coordinated. Too practiced. Everything about this screamed Taliban.

Hours earlier, a suicide bomber had detonated 1,500 pounds of explosives at Afghanistan's Supreme Court, turning buses full of civilians into twisted metal and charred remains. Seventeen dead. Forty wounded. The smell of smoke was still in the air when we'd left base, driving past the scorch marks that blackened the pavement, past the smell of burned flesh that I'd never forget.

And now we were here. Lost. In the dark. At a checkpoint that shouldn't exist.

"What the hell is wrong with these guys," my passenger muttered, his patience finally exhausted. General Washington—Army combat veteran— leaned forward in his seat. "I'll open the damn gate myself."

The unlock button.

Click.

No.

His hand moved to the door's deadbolt, fingers grasping the metal lever.

No, no, no.

The deadbolt was the only real barrier between us and them. The only thing keeping us safe. And he was voluntarily removing it.

My mind flashed to my son's face. Jordan. The promise I'd made before deploying: I'll be back before Christmas. I promise, buddy.

The door cracked open.

The three men exploded into action.

One lunged forward and wrenched the door wider. The other two grabbed General Washington's right arm, hauling him toward the opening like predators pulling prey from a den. His boots scraped against the floor mat, his face contorting with terror as he realized—too late—what he'd done.

"Get us the fuck out of here, John!" Washington screamed.

My hand shot to the 9mm on my hip. I drew the weapon, chambered a round—chak-chak—and searched frantically for a clear shot. But Washington filled the entire doorway, a writhing mass of limbs locked in desperate combat. I couldn't fire without hitting him.

Then Washington twisted with violent force, his boot connecting with one attacker's groin, his fist crashing into another's temple. The blows knocked two of them backward, giving him the seconds he needed to grab the doorframe and haul himself back inside.

But the third man kept his grip on the door, pulling with everything he had.

"Go, go, go!" Washington roared.

I knew exactly what I had to do.

There was no time for hesitation, no room for second-guessing, no opportunity to weigh options or consider consequences.

Action was survival.

Chapter 1: Threshold

The cold at Manas Air Base didn't just bite—it carved into exposed skin like a dull blade, the kind of cold that made your teeth ache and your lungs burn with each breath. Early spring in Kyrgyzstan was a lie. The calendar promised warmth, but the wind howling across the flight line had other plans. I stood on the tarmac at 0430 hours, watching the ground crew prep the C-130 Hercules cargo airplane that would carry me south into the Hindu Kush. Above, the sky hung thick and black, moonless, the type of darkness that swallows hope if you stare at it too long.

The Hercules sat there like a great metal beast, engines cycling up with that deep, rumbling growl that had become one of the soundtracks to my military career. The noise vibrated through the concrete, up through my boots, settling into my chest cavity where it mixed with everything else I was carrying. Thirteen years in uniform, and that sound still made something in my gut clench—not fear exactly, but recognition. The body remembers what the mind tries to forget.

Around me, the transit area buzzed with controlled chaos. Troops in camouflage moved in clusters, their breath forming clouds in the frigid air. Everyone had that same look—the thousand-yard stare of people caught between two worlds, no longer home but not yet at war. You could feel it in the silence between jokes; in the way hands lingered on straps and zippers as if checking them twice might change what waited ahead. I knew that look from the inside. I'd been living between worlds for several years now, ever since my life started falling apart.

Standing in pre-dawn subfreezing weather wasn't new to me. My body had learned that discipline on a Vermont farm long before the Air Force ever issued me a uniform. You move. You work. Cold is irrelevant.

Fear though—that was something you couldn't muscle through.

The Air Force had taken me across the world on different missions, but this was my first combat deployment. Standing there on that flight line, I knew this tour of duty would be unlike anything I'd ever experienced. The weight pressing down on my shoulders wasn't just the sixty pounds of gear strapped to my frame. It was the accumulated mass of the last three years— a son's seizures, a daughter ripped from my arms, a marriage that collapsed like a controlled demolition, leaving nothing but rubble and the question of who I was supposed to be now.

Major John Spargur, United States Air Force, assigned to Headquarters International Security Assistance Force. Kabul, Afghanistan. That's what my NATO orders said. That's who I was supposed to be in Kabul. But standing in that pre-dawn darkness, I felt more like a man made of scar tissue and duct tape, held together by military discipline and the desperate need to prove I could still function when almost everything that mattered had been stripped away.

"All personnel for Kabul, begin boarding procedures. Move to the rear, secure your gear!" The Loadmaster's voice cut through the flight line noise, flat and professional. No ceremony. No fanfare. Just another rotation, another batch of bodies heading into the meat grinder.

I shouldered my assault pack and joined the line shuffling toward the aircraft's rear ramp. The hydraulics hissed as the ramp yawned open, revealing the Hercules's cavernous interior—red cargo webbing, pallets strapped down tight, the narrow canvas seats running along both sides of the fuselage. The smell hit me immediately: JP-8 jet fuel mixed with hydraulic fluid, old sweat, and that particular metallic tang that every military aircraft seems to have baked into its bones.

We filed in, a procession of the deployed. I found a spot midway down the port side and dropped into the canvas seat, the webbing creaking under the combined weight of me and my gear. Around me, other passengers did the same—some young, faces still soft with that pre-combat innocence; others older, worn down by previous deployments, moving with the economy of motion that comes from doing this too many times.

The rear ramp began to close, hydraulics whining as early morning light narrowed to a sliver and then disappeared entirely. The red cargo lights seemed to pulse with the sound of the engines spooling up. I could feel the vibration building, traveling through the aircraft frame, through my seat, through my spine.

This was it. The threshold. Once those wheels left the ground, there was no going back—not for six months, not until this deployment ground itself out and spat me back home, assuming I made it back at all.

My hand moved on instinct, reaching for my wallet in the cargo pocket of my Operational Camouflage Pattern uniform. The photograph inside was creased and worn, edges soft from handling. Jordan. Four years old in the picture, though he'd just turned seven three weeks ago. His gap-toothed smile stared back at me, bright eyes full of trust that I wasn't sure I deserved anymore.

I'd taken this photo myself, three summers earlier in our backyard. He'd been playing with Jasmine—our mutt with delusions of being a guard dog—and I'd caught him mid-laugh, face lit up with that pure, uncomplicated joy that children have before the world teaches them otherwise. Before seizures. Before hospitals. Before watching his sister torn from our lives.

I'll be home before Christmas; I'd told him at Denver International Airport three days ago. We'd stood in the terminal, travelers scattered around us in various stages of their transit, and I'd crouched down to his eye level, my hands on his small shoulders. I promise, buddy. I'll be home.

He'd nodded, trying so hard to be brave, his jaw clenched the way mine does when I'm fighting to hold something back. And then he'd wrapped his thin arms around my neck and held on tight, his face pressed into my shoulder, and I'd felt his whole body shaking with silent sobs he didn't want me to see.

Don't go, Daddy, he'd whispered so quietly I almost didn't hear it. Please don't go.

But I'd gone anyway. Because that's what the orders said. Because the military doesn't care that your son has epilepsy or that you're barely keeping your head above water. Because sometimes duty and survival look exactly the same from a distance, even when you know they're not.

The seizure had come out of nowhere on a Tuesday night at a pizza restaurant in Colorado Springs, where my wife was having dinner with old friends we'd once been stationed with in Montana. I was in California, working the night shift at the Joint Space Operations Center on Vandenberg Air Force Base, when the call came in. Jordan had an uncontrollable seizure during dinner. The EMTs and doctors couldn't stop it, so he was medevac'd to Children's Hospital in Denver, where they'd placed him in a medically induced coma.

My wife's voice was high and panicked: "Come as quickly as you can."

I caught the first flight out of Santa Barbara the next morning, and by the time I reached the hospital, Jordan was still in the pediatric ICU. His tiny body was rigid, muscles locked in sustained contraction, eyes rolled back showing only whites. Status epilepticus, the doctors called it—a seizure that wouldn't stop.

They'd induced a coma to protect his brain, shutting him down like a failing computer system to prevent permanent damage. For three days, my son lay in that hospital bed surrounded by monitors and IV lines and machines that breathed for him, while neurologists ran test after test searching for a cause they never found.

He'd woken up eventually. But the seizures kept coming—sometimes weekly, sometimes daily, each one stealing a little more of the bright, energetic toddler he'd been. We cycled through medications, tried different combinations, consulted specialists across three states. Nothing stopped them completely. Nothing gave us our son back.

If Jordan's epilepsy had been the first fracture, Jayden's arrival was the moment we tried to rebuild. She'd come to us as a foster placement when she was six weeks old—a preemie, born at thirty-two weeks to a mother drowning in addiction. Three pounds of fierce survival wrapped in pink hospital blankets. We'd taken her home, fed her every two hours around the clock, watched her fight her way from fragile to thriving.

She'd become ours in every way that mattered. My son had adored his little sister. My wife absolutely loved caring for this helpless infant who needed us. For eighteen months, we'd been a family again—scarred and struggling, but together.

Then, just a week before the adoption paperwork was scheduled to be signed, my wife's mind shattered like glass under a hammer.

The paranoia had probably been building for months, symptoms I'd missed or rationalized away because I didn't want to see them. But when it broke fully into the open, it was sudden and devastating. She became convinced our family was in danger and we needed to leave the country. That the government was monitoring our house. That I was part of some vast conspiracy against her.

A schizophrenia spectrum disorder, the psychiatrist told me after the involuntary commitment. A chronic condition requiring lifelong medication and management. Treatable, but not curable.

The social worker's visit came two weeks after the diagnosis. She'd sat in our living room, clipboard on her lap, and delivered the verdict with bureaucratic efficiency: "Mr. Spargur, given your wife's mental health status and inability to consistently comply with treatment, we cannot in good conscience leave a vulnerable infant in this home environment."

"She's our daughter," I'd said, hearing the desperation in my own voice. "We've been fostering her almost since she was born. You can't just—"

"We can, and we are." The social worker's face had been professionally sympathetic but immovable. "However, we know you are a very capable father and commend you for your efforts during this difficult time. If you were to divorce your wife, we would allow you to adopt your daughter."

Divorce or lose Jayden. Those were my options. Leave my wife while she was drowning in mental illness, or let the state take our daughter.

I'd chosen option three: fight. For three months, I filed appeals, attended every hearing, presented every possible plan to keep our family intact. I documented medication schedules. Arranged psychiatric care. Proved I could manage everything. I fought with everything I had.

And on a cold Tuesday in November, standing in the lobby of social services with my son's small hand clutched in mine, I'd lost.

The C-130 lifted off, nose pitching up sharply as we clawed for altitude. My stomach dropped, that familiar sensation of leaving the earth behind. Around me, the cargo bay had gone quiet except for the relentless roar of the engines. Everyone lost in their own thoughts, their own reasons for being on this flight.

I was back in that lobby. Fluorescent lights buzzing overhead. Linoleum floors polished to a harsh gleam. The social worker approaching with professional sympathy painted on her face and Jayden in her arms.

Jayden had known something was wrong. At two years old, she'd already developed that primal instinct for danger. When I'd reached for her, she'd clung to my shirt with both fists, her face pressed against my chest, her whole-body rigid with fear.

"Say goodbye to Daddy," the social worker had said gently, prying her loose.

The sound Jayden made when they separated us from each other wasn't quite a scream—it was something more primal, a sound of pure animal terror and betrayal. She'd reached back for me, her small hands grasping at empty air, her voice breaking on a single word repeated over and over: "Daddy! Daddy! Daddy!"

I'd stood there frozen, my arms still shaped like they were holding her and watched them carry my daughter down the hallway toward a future I'd never be part of. The sound of her crying had echoed off the walls long after they'd turned the corner. Long after the door at the end of the hall had clicked shut. Long after the silence that followed felt like death itself.

Jordan had tugged at my hand. When I'd looked down, his face was wet with tears, his eyes wide with confusion and fear that no five-year-old should have to understand.

"Why are they taking her, Daddy?"

I'd dropped to my knees right there on that polished floor and pulled him into my arms, holding him while we both cried. Because I didn't have an answer. Because there was no good reason. Because sometimes love isn't enough and the world takes what it wants regardless of how hard you fight.

That was the last time we saw Jayden.

The C-130 banked hard left as we began our combat descent into North Kabul International Airport (NKAIA), the NATO base adjacent to Kabul's civilian airport. The aircraft fell from the sky in a controlled plummet designed to minimize our exposure to ground fire. My stomach lurched into my throat. The engines screamed. The airframe groaned and shuddered like a living thing as we dropped through thirty thousand feet of Afghan airspace at a rate that would have made any civilian pilot lose their license.

"Better get used to these combat landings, gentlemen!" the Loadmaster shouted over the engine roar. "The Taliban love making these big birds their targets. Can't be too careful!"

The ground rushed up to meet us. Through the porthole, I caught glimpses of brown earth, mountains in the distance, the sprawling complex of Kabul's international airbase coming into focus far too quickly.

We leveled out at the last possible second, the pilots fighting physics and momentum as they lined us up with the runway. The wheels hit concrete with a bone-jarring impact that rattled my teeth and sent a shockwave through the fuselage.

The rear ramp began to lower, hydraulics whining as harsh Afghan sunlight flooded into the aircraft's belly. With it came the heat—dry and shocking after Kyrgyzstan's cold, carrying the scent of jet fuel and dust and something else, something ancient and unamenable that rose from the Hindu Kush mountains surrounding the base.

"Welcome to scenic Afghanistan, folks!" the Loadmaster announced cheerfully. "Please exit in an orderly fashion and try not to get shot on your first day. Command frowns on that sort of thing!"

I stepped off the ramp onto Afghan soil, feeling the solid ground beneath my feet for the first time in hours. Around me, the other passengers were dispersing, heading toward whatever came next in their individual wars.

I stood there for a moment, assault pack heavy on my shoulders, weapon slung across my chest, surrounded by the sights and sounds and smells of a combat zone I'd agreed to call home for the next six months.

This was the threshold. I'd crossed it. There was no going back now— not until this deployment ground itself out and sent me home, assuming home was still there when I got back. Assuming Jordan was okay. Assuming my mom could handle it. Assuming I could survive living in a war zone while my son lived in a different kind of war back in Colorado.

Major Spargur, reporting for duty at Headquarters International Security Assistance Force. An Air Force officer assigned to a NATO billet, surrounded by coalition partners from twenty-eight nations, tasked with bringing stability to a country that had been at war ever since its conception.

But beneath the rank and the uniform and the mission briefings that would come, I was just a father—broken, bleeding, barely holding together—trying to survive long enough to make it home to the only thing that still mattered.

I stared out at the heat rising from the tarmac, warping the distant mountains into ghosts. The graveyard of empires. Afghanistan. And I'd just walked in.

Chapter 2: First Contact

April 12, 2013

"About damn time you got here, Spargur!"

The shout punched through the diesel fumes like a drill sergeant's wake-up call. I jerked my head up, squinting against the Afghan sun, and found myself face-to-face with an Air Force Major who looked like he'd been photoshopped straight out of a Pentagon recruitment ad—all chiseled jaw, sun-bleached blond hair, and a smile that suggested he knew exactly how ridiculous this all was and loved every second of it.

"You must be Daniel," I managed, though any authority in my voice was thoroughly undermined by the fact that I was currently hunched under enough gear to stock a small military surplus store.

"That I am," he said, seizing my hand and pumping it with the enthusiasm of a man trying to draw water from a dry well. "I've been dreaming about the day my replacement showed up. Full Technicolor dreams, too. Sometimes you were riding a white horse, like some kind of cavalry savior. Sometimes a helicopter, which would've been more practical. Once, and I still can't explain this one, you were on a Vespa—baby blue, chrome details, the whole works." His grin stretched wider. "Looks like my dreams are finally becoming a reality—minus the Vespa, I see. Shame about that."

Before I could formulate a response to that particular brand of sleep-deprived humor, he'd already spun on his heel and gestured toward a tall, stocky figure in Canadian fatigues.

"This is Rod, Canadian Air Force. He'll be your right-hand man here—your wingman, your battle buddy, your emotional support officer, whatever you want to call it. Rod's got the patience of a saint and can put rounds on target like he's got a personal vendetta against the laws of physics."

Rod stepped forward with an apologetic half-smile that told me he'd given up correcting Daniel's introductions some time ago. "Howdy, eh. Don't listen to him—he makes me sound way more impressive than I actually am." His handshake was firm but mercifully brief, the grip of a man who understood that after three days of military transport flights, every human interaction extracted energy I simply didn't have.

"And this Chuck Norris-looking guy here," Daniel barreled on, "is Bill. Former Green Beret, Vietnam vintage. Now he hangs out with us at our beautiful desert resort working as a lawyer for NATO, because apparently retiring to some beach town in California and playing golf five days a week wasn't sufficiently life-threatening for his taste."

Bill's face was a roadmap of experience—weathered, creased, the kind of leather-tough skin you only got from decades of sun and sand and not giving a damn. He cracked a smile that suggested he'd heard this exact introduction roughly seven hundred times and somehow still found it amusing. "The golf courses back home don't shoot back," he drawled. "Where's the sport in that?"

We loaded my gear into a tan, up-armored Toyota Land Cruiser that had seen better days—maybe better years. The armor plating added God knows how many pounds to what was already a vehicle designed more for durability than comfort.

"Kabul's a bit of a maze," Daniel said, sliding behind the wheel. "You'll want to keep a close eye on the local roads—they're full of potholes big enough to swallow a small car, and the trash you see scattered everywhere isn't just litter. Sometimes it's camouflage for an improvised explosive device (IED). Sometimes it's just trash. The trick is figuring out which is which before it figures you out."

As we rolled out through the security gates, leaving behind the relative sanity of concrete barriers and guard towers, I braced myself for what Daniel had called "the drive." The transition was immediate and jarring, like someone had flipped a switch from order to chaos.

The streets didn't just come alive—they exploded into a full-scale assault on every sense I possessed.

Mules, their ribs showing through mangy coats, pulled carts piled impossibly high with everything from firewood to rusted scrap metal, lurching into traffic without warning. Motorcycles—some carrying entire families of four or five, children sandwiched between parents like human cargo, nobody wearing helmets—wove through the mess with a reckless confidence that bordered on suicidal.

And the white Toyota Corollas. Holy smokes, the Corollas. They were everywhere, a relentless sea of them, each one looking like it had survived at least three wars and maybe a natural disaster or two. Held together by prayers, duct tape, and what I could only assume was sheer stubborn will, they dominated the roads like cockroaches—unkillable, unstoppable, and multiplying faster than you could count them.

Pedestrians didn't navigate this chaos so much as gamble with it, darting between vehicles with the fatalism of people who'd long since made peace with their mortality. Old men in traditional shalwar kameez strolled across four lanes of traffic like they were taking a Sunday walk in the park. Women in burqas, completely shrouded in faded blue fabric, stood in the middle of the street with outstretched hands, begging while traffic flowed around them like water around stones.

There were no stop signs. No street signs. No traffic lights. Just an unspoken understanding that the boldest driver won, that hesitation was weakness, and that lanes were merely suggestions for the timid.

The noise was a physical thing, an assault that made my ears ring. Horns blared in angry, discordant symphonies—not the polite "excuse me" honks of American highways, but aggressive, sustained blasts that meant "move or get hit." And then, cutting through it all, the occasional backfire—sharp, sudden, unmistakably gunshot-like—that sent my hand instinctively toward my weapon before my brain could catch up and remind me it was probably just a badly tuned carburetor. Probably.

On either side of the road, buildings stood like wounded soldiers refusing to fall. Bullet-pocked walls told stories written in concrete and blood. Collapsed roofs showed the skeletal remains of what had once been second floors, third floors, whole lives. The facades were living history books, their pockmarks and scars recording decades of conflict with brutal honesty. Soviet shells from the eighties. Mujahideen rockets from the nineties. Taliban mortars from the early 2000s. Each layer of damage added to the last, creating a palimpsest of violence that nobody bothered to erase because what would be the point? The next war would just add its own chapter.

And yet, impossibly, life persisted. Small shops somehow thrived in this chaos, their shelves crammed with an improbable mix of Chinese electronics, Russian army surplus, bootleg DVDs, bags of rice, motor oil, and things I couldn't identify and wasn't sure I wanted to. Shopkeepers sat in doorways, smoking and watching the traffic with the detached interest of men who'd seen it all and been surprised by none of it.

Children darted in and out of traffic like they were playing a particularly dangerous game of tag, their faces dirty, their clothes torn, their eyes sharp with the kind of street wisdom that came from growing up in a war zone. Some sold packets of tissues or cheap sunglasses. Others just seemed to be there, existing in the margins, part of the landscape.

"Keep your eyes peeled," Daniel said, his voice cutting through my sensory overload. He swerved expertly around a donkey cart that had stopped dead in our lane. "You'll see a little bit of everything out here. Street vendors, beggars, kids who'll steal your wallet while smiling at you. Just remember to stay vigilant. Complacency kills faster than bullets in this place."

Finally, after what felt like both an eternity and no time at all, we pulled up to the gates of the Headquarters International Security Assistance Force (HQ ISAF). The transition was jarring, almost violent in its abruptness. One moment we were in the chaos of Kabul's streets, the next we were entering the Green Zone—a secure area surrounded by thick concrete barriers, guard towers bristling with weapons, and enough razor wire to slice the sun.

The contrast was immediate. The chaos of Kabul's streets was replaced by the ordered urgency of a military installation. Everyone moved with purpose, with direction, with a sense that their time mattered and was accounted for. The familiar rhythms of military life—the flags, the uniforms, the sound of boots on pavement—provided an odd comfort after the sensory overload of the drive.

Daniel parked the Land Cruiser, killed the engine, and turned to face me. His grin was equal parts welcome and warning. "Welcome to your new life, Spargur. Let's get you settled in and figure out how to keep you alive long enough to enjoy it."

Chapter 3: Baptism by Fire

April 13, 2013

The fluorescent lights of the Dining Facility (DFAC) hummed overhead as I navigated through the morning crowd with my breakfast tray. The DFAC was already packed with soldiers from a dozen different nations, their various uniforms creating a patchwork of camouflage patterns. Conversations in Spanish, French, German, and languages I couldn't identify overlapped each other.

At six-foot-five, I had a clear view over most of the crowd, though my height made weaving between tables more challenging. My slender frame meant I could slip through gaps that would've stopped a bulkier soldier, but the low-hanging lights and doorframes seemed designed for someone a foot shorter. I'd already learned to duck instinctively after clipping my head on a support beam my first day here.

"How'd you sleep?" Daniel asked, sliding into the seat next to me with his own tray. His eyes had that particular look I'd seen in other deployed soldiers—alert even when they should be tired, always scanning.

I set down my tray and took a long sip of coffee before answering. "I've never slept in a tin can before," I said, thinking of the metal shipping container that was now my home. The Conex I shared with three other roommates was functional, sure, but comfortable wasn't a word I'd use to describe it. "Took awhile to get used to all the noises, but I finally managed to get to sleep. Everything echoes in that thing. I think I'd rather be in a tent."

Daniel smiled, the type of knowing smile that comes from experience. "Yeah, not the most comfortable living quarters. But trust me, when the bullets start flying, you'll be glad to have metal walls."

The casualness of his statement hit me harder than if he'd been dramatic about it. When the bullets start flying. Not if. When.

"You'll get used to it," he said. "The sounds, the schedule, the whole thing. Give it a week or two and that Conex will feel like home. Well, as much as any metal box can, anyway."

"How long until you head out?" I asked.

"One week. Enough time to get you up to speed on everything." He pulled out a small notebook from his cargo pocket, the pages dog-eared and filled with months of notes. "We'll start today after breakfast. I'll walk you through the ops schedule, introduce you to the key players, and show you where everything is."

"I know your orders said you'd be working as an exec for the ISAF Deputy Chief of Staff for Resources," Daniel continued, "but he's a German 2-star general and brought his own staff from home with him. Your NATO job has been loaned out to the Basing Office. So instead of reporting to the 2-star, you'll report to the Chief of Basing, US Navy Captain Landry."

I set down my fork, processing what he'd just said. My orders had been clear—executive officer to the ISAF Deputy Chief of Staff for Resources. That's what I'd prepared for, what I'd expected. And now, twenty-four hours after landing in Kabul, the mission had changed entirely.

Basing operations.

I had worked a lot of different jobs during my Air Force career, from nuclear Intercontinental Ballistic Missiles in Montana to tracking enemy satellites in the depths of NORAD's Cheyenne Mountain in Colorado, but when it came to basing operations... I was clueless.

"Don't worry, I didn't know anything about basing either before I arrived," said Daniel, who must have seen the look of concern on my face. "It's going to be a bit of drinking from the firehose, but I'll teach you everything you need to know."

We walked past the HQ ISAF building where the Commander of all NATO forces in Afghanistan, Joseph Dunford, a four-star Marine general, carried out his daily duties. The building loomed large both physically and symbolically—the nerve center of the entire Afghan operation.

From the HQ, we took a right and walked to a two-story building, heading up the stairs that ran along the outside wall. The metal steps clanged under our boots, each footfall announcing our approach.

Daniel knocked on a door on the left. "Let me introduce you to General Washington. He's the Chief of Combined Joint Engineering and our basing office falls under his division."

"Enter."

Sitting behind a small desk was a battle-hardened Army soldier. General Washington looked exactly like what you'd expect from someone who had served multiple combat deployments and was nearing completion of his first tour in Afghanistan. His uniform was crisp, and his eyes carried that same alertness—the look of someone who'd learned to stay sharp because their life depended on it.

"Sir, this is Major Spargur," Daniel said. "He's my replacement in the Basing Office."

General Washington stood and extended his hand across the desk. His grip was firm, deliberate. "Welcome to Combined Joint Engineering command. Daniel's been doing good work for us. Hope you're ready to hit the ground running."

"Yes, sir. Looking forward to it."

Washington gestured for us to sit in the two chairs facing his desk and began laying out the mission.

"Let me give you the lay of the land organizationally," he began. "At the top, you've got Commander ISAF—he's running the whole show in Afghanistan. Helping him out is his Chief of Staff, who manages the planning, administration, and coordination of his directives. There are several staffing offices under the Chief of Staff that support General Dunford, like mine, HQ Engineering. I am the Chief of Engineering. The Basing Office, where you'll be working, is a branch of my office. Then, you have the subordinate Headquarters, ISAF Joint Command (IJC), who is tasked to take care of the day-to-day tactical operations. Their HQ is at NKAIA, where you landed yesterday. Under IJC, there are six regional commands: RC Capital, RC West, RC East, RC North, RC South, RC Southwest.

He leaned forward slightly, his expression growing more serious. "As Chief of Engineering, we oversee everything related to building, maintaining, and protecting the infrastructure that keeps this mission running. We manage hundreds of construction projects across the country, from forward operating bases to roads that allow our forces to move safely. We coordinate with Afghan National Army engineers to build their capacity."

He paused, letting that sink in before continuing. "During the height of the surge in 2011, ISAF had over 130,000 troops at over 800 bases across the country. By this summer though, ISAF's mission will transition from leading combat operations to a train, advise, and assist role. By the end of 2014, our mission here will be done and we'll be handing the keys completely over to the Afghans."

"As you can tell, there's not much time left to figure out how to close or transfer the remaining 346 bases still open," Washington continued. "Figuring out how to do this quickly and safely is a top priority for General Dunford."

He leaned forward now, his elbows on the desk, his gaze fixed squarely on me. "Now it's on the Basing Office's shoulders to get this done."

The enormity of what he'd just described settled over me like a mountain of files, each one critical, all of them due yesterday. Over three hundred bases. Less than two years. And I had one week to learn a job I'd never done before in order to help make it happen.

"Alright gentlemen, I've got a meeting to run to," Washington said, standing. "I'll let Captain Landry and the Major fill you in on the roles and responsibilities of your job." He extended his hand again. "Welcome to Combined Joint Engineering, Major Spargur. Now get to work."

He said it with a smile, but there was a serious undertone that made it clear this wasn't a casual suggestion. This was the mission, and the clock was already ticking.

We crossed the hall to the Basing Office. Daniel knocked once and pushed the door open without waiting for a response. Inside, a lean Navy officer with greying black hair looked up from his desk, his face breaking into a welcoming smile.

"Dan!" he said, standing quickly. "This must be your replacement."

"Captain Landry, this is Major Spargur," Daniel said.

Captain Landry was a former P-3 reconnaissance pilot, his demeanor carrying earnest, slightly nervous energy combined with an eagerness to please. He had an enthusiastic quality, and I suspected beneath it was a sharp mind that had managed countless reconnaissance missions over hostile territory.

"Welcome, welcome!" Landry said, extending his hand and shaking mine vigorously. "We're glad to have you. Daniel's been invaluable, and I know you've got big shoes to fill, but we'll get you up to speed in no time."

He gestured to a small desk wedged against the wall near his own. "That'll be your workspace. Not much, but it's functional."

I took a seat at what would become my desk while Daniel settled into a chair beside me. Captain Landry pulled out a laminated chart and spread it across his desk.

"First things first," he said. "You need to understand how we categorize bases here. Everything flows from this."

He pointed to the top tier of the chart. "Strategic bases are your heavy hitters. They hold ten thousand troops or more. These have airports, hospitals, the works. They're your logistical hubs, your intelligence centers, your command-and-control nodes. Think Bagram and Kandahar."

His finger moved down. "Operational bases sit in the middle—four to eight thousand troops. They've got medical facilities, and substantial infrastructure. Their job is to connect the strategic bases with the tactical forces out in the field."

"And then you've got your Tactical bases," Landry continued, his voice taking on a more serious tone. "These are the lowest level, focused on direct combat support. No airports. Limited medical facilities. Just enough to keep soldiers alive and fighting."

"Now here's what matters for your job," Landry said, leaning back. "At ISAF headquarters, we manage the transfer and closure of strategic and operational bases. That's our lane. For the tactical bases—all those small combat outposts and patrol bases scattered across the country—we provide guidance and oversight, but the actual execution falls to our subordinate headquarters, IJC. They handle the tactical stuff and portions of the operational level."

Daniel spoke up. "It's a cleaner division of labor than it sounds. We deal with the big, complex bases that require ministerial approval and international coordination. IJC deals with the small outposts that are often just handed to the local Afghan National Army unit."

"Which brings us to the heart of what we do—the ISAF Theater Basing Roadmap," Landry said, pulling out another document with milestones and timelines. It looked like a project management spreadsheet that had been put on steroids and sent to war.

"This roadmap lists every base in the theater," Landry explained. "Their projected date of transfer or closure, the major milestones needed to get them to their end state, who's responsible for what. Basically, this is the phased plan to restructure, shrink, and complete the handoff or closure of all coalition and ISAF bases in Afghanistan."

I stared at the document. Each line represented months of planning, millions of dollars, thousands of lives affected.

"As we shift from combat operations toward train, advise, and assist," Landry went on, "every decision we make has to support that transition. We're not just closing bases. We're managing the end of a war."

The significance of that statement hung in the small office for a moment.

"Part of that management involves weekly meetings with deputy ministers from the Government of the Islamic Republic of Afghanistan—GIRoA. There are around twenty-five different ministries, but only a handful have any interest in acquiring the bases we're closing. Typically, we're dealing with Agriculture, Irrigation and Livestock, Commerce and Industry, Defense, Finance, and Interior Affairs."

"Actually," Landry said, checking his watch, "our next meeting with them is later today. We're discussing the transfer of an operational base called Warehouse. You'll tag along with Dan. Watch, listen, take notes."

Daniel caught my eye and gave a small nod. I'd be shadowing him, learning on the fly.

"Here's the thing though," Landry said, and his tone shifted slightly. "I'm leaving in a few weeks for a NATO operational planning working group in Belgium. I'll be gone for about a month." He let that sink in before continuing. "Pay close attention to everything Dan teaches you over the next few days before he departs. Because once he's gone and I'm still in Europe..."

He didn't need to finish the sentence. I understood.

"You'll be running the place until I get back."

The fluorescent lights hummed overhead. Outside, I could hear the sound of a Blackhawk helicopter taking off. Somewhere in this compound, General Dunford was making decisions that would shape the end of this war. And in less than two weeks, I'd be the one managing a critical piece of that puzzle, with barely enough time to learn the basics.

"Yes, sir," I said.

It was the only response that made sense. There was no room for doubt, no space for admitting I felt completely unprepared. This was the mission. I'd figure it out because there was no alternative.

Daniel walked me through the rest of the office, introducing me to the team that would become my lifeline over the coming months. Rod would be my wingman. Bill, the former Green Beret turned NATO lawyer, dealt with land disputes. Stephanie, the Canadian environmental expert, made sure we weren't leaving behind toxic waste dumps. The Polish engineers, Dombroski and Nowak, handled structural assessments. Jules, the British RAF officer, managed airfield certifications.

Each of them had their specialty, their piece of the puzzle. Together, we were supposed to manage the orderly withdrawal of an empire.

After lunch, Daniel led me back outside into the midday heat. "Come on," he said. "Time to show you the routes."

We walked across the compound toward the motor pool. A handful of Toyota Land Cruisers sat in a designated area, distinct from the rows of armored MRAPs and Humvees. I recognized the vehicle immediately—Daniel had picked me up from the airport in one just yesterday, though in the chaos of arriving, I hadn't paid much attention to it.

"This one is ours," Daniel said, pulling out a set of keys and patting the hood of the closest Land Cruiser. "One of the perks of the job."

I ran my hand along the door frame. The vehicle looked civilian enough—no gun turret, no obvious armor plating—but Daniel tapped the window with his knuckle. "Bulletproof glass," he said. "And there are steel plates hidden in the doors and undercarriage. Not nearly as robust as an MRAP, but better than nothing. These things are modified to blend in while giving us at least some protection."

I nodded, feeling slightly better about the setup, though not by much. Better than nothing wasn't exactly a ringing endorsement when it came to surviving an IED or ambush.

"Ninety-eight percent of the people on this base," Daniel continued, climbing into the driver's seat, "they can't go anywhere without an armored convoy. Minimum two vehicles, fully armed escort, planned routes, timed movements. The whole nine yards." He started the engine, and it turned over with a reassuring purr. "But because of our unique mission—because we're coordinating with Afghan ministries, visiting bases across the country, meeting with local officials—we've got our own vehicle. No convoys required."

I settled into the passenger seat, my long legs cramped even in the spacious interior. "And we can just... go anywhere?"

"Anywhere in Afghanistan, anytime we want." Daniel said it with a grin, like he was describing an all-access backstage pass. "No waiting for convoy schedules. No coordinating with movement control. We just get in and go. Makes our job a hell of a lot easier."

He pulled out of the parking area and headed toward the main gate. The guards waved us through without hesitation, clearly familiar with our vehicle and mission.

As we drove through the outer perimeter and onto Kabul's chaotic streets, Daniel kept talking, pointing out landmarks and explaining routes. But I wasn't really listening anymore. I was watching the streets, the people, the other vehicles.

Every car that got too close felt like a threat. Every intersection was a potential ambush point. Every pedestrian on the sidewalk could be carrying anything under their loose clothing. The sun glinted off the Land Cruiser's unarmored windows, and I found myself calculating angles of fire, blast radiuses, escape routes.

Daniel was describing this freedom of movement like it was a privilege, a convenience that would make our jobs easier. And maybe it was. But all I could see were the dangers. We weren't just outside the wire—we were outside the entire security apparatus that kept most soldiers alive. No gun trucks. No emergency response teams tracking our movements.

Just us, alone, in a civilian vehicle that might as well have had a target painted on the side.

"You'll get used to it," Daniel said, misreading my silence. "After a few trips, you won't even think twice about it."

I nodded but kept my mouth shut. My more reserved nature meant I tended to process things internally anyway, but right now I was glad for the excuse not to voice what I was thinking. Because what I was thinking was that this wasn't a perk of the job. This was Russian roulette with extra steps, and I was going to be playing it multiple times every week from here on out.

We passed a burned-out building, its walls pockmarked with bullet holes. A reminder that this city was still a war zone, no matter how quiet it seemed in this moment.

"See that?" Daniel pointed to a compound surrounded by concrete blast walls. "That's Camp Warehouse. You'll be spending a lot of time there—it's one of the operational bases we're working on transferring." He drove on, turning down a side street. "And over here is the main route to Camp Phoenix. You'll want to memorize all these roads. Makes navigation easier when you're on your own."

When you're on your own.

The words settled in my stomach like a gut punch.

I glanced at the dashboard. No GPS. No navigation system. Just the dusty windshield and Daniel's hand gestures pointing out turns and landmarks.

"How many bases am I going to need to know how to get to?" I asked.

"In the Kabul area? At least eighteen." Daniel said it casually, like he was talking about memorizing a grocery list. "Camp Phoenix, Camp Warehouse, Camp Bala Hissar—that one's up in the old fortress, great views but a pain to get to. Then there's KMTC, Camp Blackhorse, North Kabul International Airport..." He rattled off names while simultaneously navigating through an intersection clogged with civilian traffic. "You'll memorize them quick enough. After a few trips, it becomes muscle memory."

Daniel continued the tour, pointing out safe houses, checkpoints, and areas to avoid. He talked about blast patterns and standoff distances, about reading traffic for suspicious behavior, about what to do if we took fire. He made it all sound routine, just another part of the job.

But every word reinforced what I already knew: this wasn't routine. This was calculated risk; every single time we left the gate. And soon, I'd be the one making those calculations, deciding whether a meeting was worth the exposure, whether a site visit was worth the vulnerability.

We turned back toward the ISAF compound, passing through a marketplace where vendors hawked vegetables and electronics side by side. Normal life, continuing despite the war. Or maybe because of it.

As the gate guards processed us back onto the base, I felt the tension in my shoulders ease slightly. Back inside the wire. Back to relative safety.

Daniel parked the Land Cruiser in its designated spot among the other Land Cruisers at the motor pool. "Not bad for your second ride in one of these, right?" He climbed out, completely at ease. "Give it a week. This'll all feel normal."

I followed him out of the vehicle, my height making me unfold from the passenger seat. I looked back at the tan Toyota, sitting there with its hidden armor and bulletproof glass among the obvious military behemoths. To Daniel, it represented freedom and efficiency. To me, it looked like a calculated risk I'd be taking every single day—only now with the added pressure of navigating a city I didn't know, to bases I'd never seen, without so much as a GPS to guide me.

But I kept that thought to myself. I nodded to Daniel, said something noncommittal about appreciating the orientation, and followed him back toward our office.

As we stepped back into the cramped workspace, the weight of everything I'd just learned settled onto my shoulders like a rucksack full of stones. Three hundred forty-six bases. Weekly meetings with Afghan ministers. A roadmap I didn't understand managing a drawdown I hadn't planned. And now, a vehicle that gave us freedom to travel anywhere— which really meant freedom to be vulnerable anywhere.

The morning sun streamed through the windows at the end of the hall, catching the dust particles that seemed to hang perpetually in the Afghan air. From somewhere outside came the steady rhythm of construction—

hammering, buzzing, the backup beep of heavy machinery. The sound of a war winding down, one base at a time.

Except now, I was the one who had to help make sure it actually wound down instead of just grinding to a halt. And I'd have to do it while driving through a war zone in an armored civilian vehicle, trying not to get lost and pretending it was just another day at the office.

Chapter 4: Outside the Wire

My stomach sank the moment I stepped outside the gate of HQ ISAF. It wasn't a gradual thing, a slow dawning realization. It was immediate, visceral—like stepping off a cliff you didn't know was there.

One foot inside the Green Zone, one foot out. That's all it took to feel the difference.

Inside—concrete barriers thick enough to stop a truck bomb, armed guards at every checkpoint, the carefully maintained illusion of control. Outside—open street, open sky, open season. Exposure that made your skin crawl and your peripheral vision go into overdrive, scanning for threats that could materialize from any direction, at any moment.

Daniel was back stateside by now, probably sleeping in a real bed and eating food that didn't come in a tan plastic pouch. Captain Landry was on Temporary Duty in Belgium, attending meetings in climate-controlled rooms where the biggest threat was PowerPoint-induced boredom. Which left me—new guy, still learning where the latrines were—standing on a

Kabul street corner about to run a weekly Basing meeting with Afghan ministry representatives, pretending I knew what the hell I was doing.

"Tell me again why we don't wear our helmets and body armor outside the wire to meet these guys?" I asked my Canadian coworker, Rod, already knowing the answer wouldn't make me feel any better.

Rod glanced at me with the patient look of a man who'd answered this question before. "Wearing all our battle rattle doesn't foster an environment of trust with the Afghans, or something like that, eh." His thick Newfoundland accent made even the bureaucratic jargon sound almost folksy. "They show up here in their suits, looking professional, and if we're dressed like we're going to war—helmets, plates, the whole kit and kaboodle—it affects our ability to develop close relationships with them."

"Makes sense, I guess," I said. My mouth said one thing. Every nerve ending in my body screamed something else entirely.

I knew the story. Everyone knew the story. A few months before I arrived, a suicide bomber had approached this exact spot—this exact stretch of cracked pavement where I was now standing. The bomber had targeted the Afghan ministry reps and the NATO soldier escorting them onto base, choosing the moment of maximum vulnerability when they were exposed, outside the wire, unprotected.

It was the neighborhood kids who saved them. Kids who'd grown up in a war zone, who could read danger the way other children read picture books. They recognized a face that didn't belong, saw something hidden under a coat on a warm day, understood the geometry of death before most of them could spell their own names. They screamed. They ran. They created the chaos that broke the bomber's plan.

The Afghans dove behind their van. The NATO soldier ducked behind a concrete barrier. The bomber knew his cover was blown, so he did what bombers do when cornered.

He detonated.

Hundreds of metal ball bearings—the kind designed specifically to shred human flesh—tore through the air at supersonic speeds. Two children were killed instantly, their small bodies absorbing the blast meant for others.

Everyone else walked away with what the after-action reports clinically called "minor injuries," which is military speak for "alive but never the same."

Now I stood just feet away from where it happened. Someone had cleaned most of the blood off the pavement, and if you knew where to look you could still see the scorch marks, the pockmarks in the concrete barrier where ball bearings had embedded themselves.

Within a minute of Rod and me stepping onto the street, we were swarmed.

Children. Almost a dozen of them, appearing from nowhere. Some bearing scars on their faces and arms—jagged pink lines that caught the light, shiny patches of skin that had healed wrong and tight—souvenirs from the attack months before. They were dressed in ragged clothes, many without shoes despite the rocks and broken glass littering the street.

Each one carried boxes or bags of trinkets—cheap bracelets, hard candies, wilted flowers, handmade crafts of dubious quality. They held them up like offerings to foreign gods.

"Mister, mister! You buy! Very good! Very cheap!"

I'd made a mistake the week before. A small girl, maybe five years old, had approached me with a smile—this beautiful, radiant smile that cut through the dirt caked on her cheeks like sunlight breaking through storm clouds. She held up a handmade bracelet, crude but colorful. Her eyes met mine, and in that moment, she wasn't a street kid or a beggar or part of some larger geopolitical tragedy. She was just a little girl who reminded me of the daughter I had lost, holding out something she'd made, hoping I'd buy it.

So, I gave her five Euros.

The second I waved off the other children, chaos erupted. The older kids descended on her like wolves. Just knocked her down, no hesitation, no mercy—shoved her flat onto the pavement. The money flew from her hand. A fight broke out immediately, bodies piling on, small fists swinging for the crumpled Euro. The little girl disappeared under the mass of them, and I could hear her crying.

Daniel had been with me that day. He barked something in Dari that cut through the noise like a whip crack. The effect was immediate. They all stopped. The little girl sat up, crying, her money gone, her bracelet crushed under someone's foot.

"Hope you got more cash, John," Daniel had said, grinning. "Because now you're going to buy something from each one of them. Otherwise, they'll tear themselves apart right here on this street, and that'll be on you."

Seventy Euros later, I had eight bracelets I'd never wear, four pieces of candy I didn't want, and a wildflower that was already dying.

"You're going to go broke if you do this every week," Daniel had told me on the walk back to base. "I know you want to help these kids out— trust me, I get it, we all do—but they're pretty much all orphans. The war took their parents, disease, violence, bad luck, whatever. Now they work for gangs. Criminal enterprises use them to beg for money, to steal, to smuggle, whatever needs doing. The proceeds go to the gang leaders, not the kids. So, you're better off not supporting criminal activity, even indirectly. You want to help? Give them food or candy instead. Something they can use themselves."

Today I came prepared. Rod and I had raided the base market and filled our pockets with bags of M&Ms—the small individual packs they sold by the box. We handed them out like it was Halloween back home, tiny packages disappearing into small hands faster than we could distribute them.

"That should keep them out of our hair for at least a few minutes," Rod said as the kids ripped open the bags with their teeth, their attention shifting away from us and toward the chocolate. Their faces transformed— momentary joy, pure and uncomplicated, simple pleasure that made you forget for just a second where you were and why.

"I'm starting to notice that Afghans seem to have their own clocks," I said, glancing down at my watch for the third time. We'd already been standing here fifteen minutes past the scheduled meeting time.

"No doubt, eh. Sometimes I've waited for an hour. Sometimes they just don't show at all." Rod scanned the street with casual alertness. "They vary their times so it doesn't become a recognizable pattern. If they showed up at exactly 1400 hours every Thursday like clockwork, the Talibs would know exactly when and where to hit them. The Talibs have eyes everywhere."

"Makes sense from their perspective," I said. "But that sucks for us. We're like sitting ducks out here."

The words were barely out of my mouth when a brown Toyota van rounded the corner and pulled up beside us. Five Afghans in Western suits exited, stepping down from the van and straightening their jackets with practiced motions. They looked impossibly neat and composed, like they'd just stepped out of an air-conditioned office building rather than driving through Kabul's anarchic streets.

"As-salamu 'alaykum," I said, approaching them with a smile I hoped looked confident.

"Wa 'alaikum salam," each one replied in turn, shaking our hands with firm grips and direct eye contact. Peace be upon you, and upon you be peace. An exchange of blessings in a place where peace was mostly theoretical.

Rod and I fell into position, leading them back toward the Green Zone, toward the foot traffic entrance of HQ ISAF. We moved at an easy pace, not rushing but not dawdling either. The walk felt longer than it was— maybe one hundred yards of exposed street where anything could happen. Every step back toward those concrete barriers was a step closer to breathing normally again.

Once we got them processed through security, we proceeded through the maze of concrete barriers and serpentine walkways designed to prevent vehicles from building up speed and made our way to the large conference room in one of the administrative buildings sitting behind the Headquarters.

Three U.S. Army soldiers who served as advisors to the Afghan ministry reps were already waiting, seated on one side of the long table. They looked tired, the kind of tired that comes from months of deployment and endless meetings that never quite accomplish what you hope they will. An Afghan interpreter—a middle-aged man in Western clothes who'd probably been threatened a hundred times for doing this job—sat at the head of the table, his notebook open and ready.

The meeting focused on wrapping up negotiations for turning over Camp Warehouse, one of the dozens of operational bases NATO was in the process of transferring back to Afghan control as part of the withdrawal. The base had significant infrastructure—buildings, utilities, power systems, water treatment, maintenance facilities, all the things that cost real money to maintain and operate.

I tried to convince the Afghans to let NATO demolish some of the more expensive structures before the handover. It seemed logical to me, practical.

"These buildings require constant maintenance," I explained, pausing after each sentence to let the interpreter translate. "Right now, you're about to inherit hundreds of new bases as NATO forces draw down. But here's the problem—once U.S. and NATO funding leaves with the withdrawal, you won't have the budget to maintain them all. Some of these structures will cost more to keep standing than they're worth."

The interpreter spoke in rapid Dari, his hands moving expressively. The Afghan officials listened politely, their faces neutral, revealing nothing. They conferred among themselves in low voices, heads tilted together. Then their senior representative—a distinguished-looking man with gray in his beard and eyes that had seen too much—responded through the interpreter.

They refused.

They wanted to keep everything. Every building. Every wire. Every structure, down to the last stick of wood and piece of rebar. The interpreter's voice stayed level, professional, but I could hear the finality in the Afghans' position. This wasn't up for discussion.

Maybe it was pride. Maybe it was the principle of the thing—taking back what was theirs, on their terms, complete and whole, refusing to accept the charity of demolition from an occupying force that had overstayed its welcome. Maybe they knew something about making do with limited resources that we Americans, with our endless budgets and disposable equipment, had forgotten.

After twenty minutes of this ritualized back-and-forth, I realized I wasn't going to win. This wasn't about logic or practicality. This was about something deeper, something that didn't translate neatly into military planning documents and cost-benefit analyses.

The meeting ended the way these meetings always ended—with handshakes and smiles and the unspoken understanding that we were all just going through the motions of something that had already been decided by people in rooms far away from here.

Rod and I escorted them back through the maze of HQ ISAF, back out through security, back to the pickup spot outside the Green Zone. We stood on the street again, waiting for their van to circle back around. The afternoon sun was lower in the sky now, throwing long shadows across the potholed pavement.

The brown Toyota rounded the corner right on schedule this time. The five men climbed in with minimal ceremony, settling into their seats. We exchanged final waves through the windows—friendly, professional, the international language of "until next week."

Then they were gone, and Rod and I walked back through the gate, back into the Green Zone, back into the illusion of safety that concrete and armed guards provided.

My body armor was waiting for me right where I'd left it. Thank God, I didn't need it that day.

Still, as I settled back into my office, a quiet realization settled in. I was learning something essential about survival in this place—not just the physical kind, though that mattered—but the deeper kind. The kind that let you keep functioning when everything around you was falling apart.

These Afghans had been doing it for decades. Adapting. Enduring. Finding ways to keep moving forward even when the ground kept shifting beneath their feet. Maybe there was something to learn from that. Maybe survival wasn't just about armor and weapons and concrete walls.

Maybe it was about something harder to quantify. The ability to bend without breaking. To hold onto what mattered while letting go of what didn't. To find small moments of humanity in the middle of inhumanity.

As I walked back to my office, I thought about those children with their M&Ms, their faces lighting up over something as simple as chocolate. About the Afghan officials who insisted on keeping every building, even the ones that would bankrupt them, because sometimes principle mattered more than practicality.

About my son back home, fighting his own battles and a world that didn't always make sense.

We were all just trying to survive, in our own ways. Some of us had concrete walls and body armor. Others had pride and stubbornness and a form of resilience that comes from generations of practice.

I was beginning to understand that maybe the real war wasn't the one being fought with bullets and bombs. Maybe it was the quieter one—the daily struggle to remain human in a place designed to strip that away from you.

And maybe, just maybe, I was learning how to fight that war too.

Chapter 5: Taking Command

May 16, 2013

I woke to the metallic taste of grit in my mouth and the familiar hum of a Blackhawk helicopter landing outside. Relief washed over me as I remembered—today was the day. Captain Landry was finally coming back from Belgium.

These past two weeks had been a brutal test of survival. Rod, Bill, and the rest of my office mates had thrown me lifelines every single day, walking me through the labyrinth of NATO basing operations one crisis at a time. But despite their patience, I felt like I was treading water in the deep end wearing combat boots. Every email brought a new problem I didn't know how to solve. Every phone call revealed another gap in my knowledge.

I needed Landry back. I needed someone who actually knew what the hell they were doing to take the wheel before I drove this whole operation into a ditch.

The chow hall was already packed when I arrived, the usual morning chaos of soldiers shuffling through the serving line with trays. I scanned the crowd for Landry's face, mentally rehearsing the list of questions I'd compiled over the last few weeks. No sign of him. Probably sleeping off the jet lag after a late flight. I'd catch him at the office.

After choking down breakfast—powdered eggs and something that might have been sausage in a previous life—I made my way to check the mailroom. Two packages sat waiting on the metal shelf. One had Bill's name scrawled across the top in marker. The other bore the return address of an elementary school in Illinois.

Every month like clockwork, the kids from that school sent us letters written in crayon and marker, telling us to "stay safe" and "be brave," along with a couple boxes of Girl Scout cookies. Those packages were better than any medal. They reminded us that somewhere beyond the blast walls and concertina wire, normal life still existed.

I tucked both packages under my arms and headed toward the office. I was reaching for the door handle to the Basing Office when a voice cracked across the hallway like a rifle shot.

"Spargur! My office. Now."

I turned to see General Washington standing in his doorway, his jaw set tight.

"Yes, sir." I pivoted on my heel and walked the ten paces to his office, my mind already racing through possible infractions.

"Close the door."

I pushed it shut behind me. Washington gestured to the chair across from his desk. I sat, and my breakfast suddenly felt like a brick of wet cement in my gut.

"Congratulations, Major Spargur." His voice was flat, businesslike. "You've been promoted to Chief of Basing. As of this moment, you're in charge of leading all basing operations for NATO in Afghanistan."

"Sir?" The word escaped before I could catch it.

Washington leaned back in his chair, his fingers steepled in front of his face. "There's been an incident with Captain Landry. During his layover in Dubai, he had one too many drinks before boarding his flight to Kabul. By the time he got on the plane, he was extremely intoxicated. So drunk, in fact, that he convinced himself the person sitting next to him was a terrorist. Told the flight attendant he'd witnessed this passenger performing some kind of ceremonial hand cleansing ritual because they were about to bomb the plane."

Holy shit.

"Dubai police got involved," Washington continued. "Fortunately for Landry, there were several NATO officers on the same flight who took charge of the situation and kept him out of a Dubai jail cell. But that's where his luck ran out."

I sat there, trying to process what I was hearing. Landry had seemed solid. Professional. Maybe a little worn down, but who wasn't after a year in this place?

"This last year must have really broken him," I said quietly.

Washington nodded, his expression grim. "I can only imagine what it felt like—being free in the outside world, having a few drinks at an airport bar, and then facing the reality of coming back to this place. Some people hit their breaking point. Landry hit his."

The room felt smaller suddenly, the air thicker. If it could happen to Landry, it could happen to anyone.

"Security forces are escorting him to base this afternoon," Washington said. "He'll sign paperwork, pick up his gear, and then we're putting him on the first flight back to the States. I need you to take two of your people and have them box up his belongings. You'll also need to take control of his weapons. Make an appointment with JAG this week to generate the transfer paperwork."

"Yes, sir. I'll get it taken care of."

Washington leaned forward, his eyes locked on mine. "One more thing. You've got a Top Secret/Special Compartmented Information clearance, right?"

"Yes, sir." I felt my pulse quicken. "Why?"

"Good. Captain Landry was the ISAF NATO liaison to the CIA. He meets with them monthly, sometimes weekly, to coordinate on matters that are above my clearance level. His next meeting was supposed to be tomorrow at 1000 hours." He paused, letting that sink in. "You're the ISAF representative now. Show up at the US Embassy Annex at ten sharp. Got it?"

Before I could respond, the world exploded.

The blast wave hit like a physical punch. The window rattled violently in its frame. The floor bucked beneath my feet. My teeth clacked together. The sound was massive, a deep concussive boom that I felt in my chest cavity, followed by the high-pitched ring of shocked silence.

We both jumped to our feet and rushed to the window, scanning the skyline for the telltale column of smoke. In the distance, black smoke was already mushrooming into the pale morning sky.

"That was a VBIED," Washington said, his voice distant, mechanical. "Vehicle-Borne Improvised Explosive Device. I've heard that sound too many times in Iraq to mistake it for anything else."

"Damn." My mouth was dry. "I hope no one got seriously hurt."

Even as I said it, I knew better. We both did. In a city as densely packed as Kabul, someone always got hurt. A lot of someone's.

"Let's hope," Washington said, but there was no conviction in his voice. He turned from the window and sat back down at his desk, his face already shifting back into command mode. The explosion was filed away. There was work to do.

"Alright," he said, pulling himself back to the business at hand. "You're going to have a lot on your plate now, and unfortunately there won't be anyone to backfill your position. You'll need to lean hard on Rod and the rest of the team. I'm going to shotgun an email out this morning to all the divisions announcing you're taking over as the new Chief of Basing. From here on out, you've got the stick. Understood?"

"Yes, sir."

"Good. Now go grab some people to take care of Landry's belongings. And don't forget about the CIA meeting tomorrow."

I walked toward the door, my legs feeling like they belonged to someone else. My brain churned through a thousand questions, trying to sort through the chaos of the last ten minutes. Chief of Basing. CIA meetings. Landry losing his mind in Dubai. A bomb somewhere in the city.

I paused at the door and took a deep breath, held it for a three-count, then slowly exhaled. This was it. Sink or swim time.

I pushed the door open and stepped into the Basing Office.

Everyone was already at their workstations, the normal hum of productivity. Rod looked up from his computer with his usual friendly smile.

"Did you get pulled into the principal's office this morning?" he said with a chuckle. "I heard talking in Washington's office and figured it must be you when you didn't show up here on time."

"Yeah, it was me." My voice came out flat, hollow.

Rod's smile faltered. "Everything alright?"

"I don't know. Too early to tell."

I moved to the center of the room, cleared my throat, and raised my voice. "Listen up, everyone. I need your attention for a couple minutes."

The typing stopped. Conversations died mid-sentence. Everyone turned to look at me.

I told them everything General Washington had just told me. The words felt surreal coming out of my mouth. By the time I finished, the shock on their faces probably matched what mine had looked like in Washington's office.

The silence hung in the air for a beat too long.

"Okay," I said, breaking through it. "If there are no questions, I need two volunteers to go pack up Captain Landry's belongings."

Rod's hand went up first. He didn't hesitate. Jules, the quiet British engineer, raised his hand second.

"Thanks. Get it done as quickly as you can."

They nodded and headed out. The rest of the office slowly returned to their work, though the energy had shifted. The air felt heavier now.

Bill looked up from his coffee, his weathered face unreadable. "Not the morning you were expecting, huh, John?"

"That's putting it mildly."

"Well, it's too bad about Landry. But he'll land on his feet. They always do." He took a sip of his coffee. "You've got big shoes to fill now. You ready?"

"I sure as hell don't feel ready. Not that I get a choice in the matter."

"Nobody ever feels ready," Bill said. "You just do it anyway."

Before I could respond, the phone on Landry's desk started ringing. I stared at it for a second, then realized—that's my phone now.

I walked over and picked up the receiver. "Basing Office, this is Major Spargur, how can I help you?"

"This is Sergeant Jones from the Ops center." The voice on the other end was tight, urgent. "I'm trying to reach Captain Landry."

"I'm Landry's replacement. Is there something I can assist you with?"

"You're the Chief of Basing, then?" There was a pause, the sound of papers rustling. "You have your own vehicle, correct?"

The tone in his voice made my stomach drop. Something was wrong. Something bad.

"Yes, that's right."

"Sir, we need your help. There's been an attack. We're getting reports of mass casualties at the blast site. We've already dispatched several teams, but they need more medical support immediately. Can you drive two medics to the attack site right now?"

The world seemed to tilt slightly. The explosion we'd heard. The smoke. This was it.

"Affirmative. What's the location?"

"The medics will be waiting for you at the motor pool in five minutes. Drive them southeast from the headquarters towards the Shah Shahid neighborhood and Jalalabad highway. It's under a mile, not too far from the Ministry of Defense building." He paused. "Look for the smoke and listen for the sirens. Good luck, sir."

The line went dead.

I stood there for a second, the phone still in my hand, my brain trying to catch up with what had just happened. In the span of thirty minutes, I'd been promoted, briefed on CIA meetings, and now I was about to drive into a fresh bombing site.

I hung up the phone and quickly began to gather my gear.

"Bill, you've got the office. I'm heading out."

He looked up sharply. "Where?"

"The blast site. They need someone to transport medics."

His face went pale. "Jesus, John. Be careful."

"I will."

I strapped on my vest, grabbed my helmet, picked up my M4 and headed for the door. My first day as Chief of Basing was about to get a whole lot more complicated.

But as I jogged toward the motor pool, something strange happened. The panic I'd felt moments earlier began to crystallize into something else. Focus. Determination. The kind of clarity that comes when you stop worrying about whether you're ready and just start doing what needs to be done.

For the first time since stepping off that C-130 in Kabul, I felt like I might actually belong here.

Chapter 6: The Crater

May 16, 2013

The motor pool was a maze of dusty vehicles parked in neat rows, the air thick with diesel fumes and the metallic tang of hot engines. Two figures stood near the gate, both geared up and clutching medical bags like lifelines—a Navy Corpsman and a French Army doctor in his fatigues, his beret folded and tucked in his cargo pants pocket.

"Are we expecting anyone else?" I called out as I jogged up to them, my body armor bouncing with each step.

"No sir, we're it," the Navy Corpsman replied. His voice was steady, professional, but I could see the tension in his jaw.

"Okay then, let's go. My vehicle is right here in the second row."

We piled into my tan Toyota Land Cruiser, the medics throwing their bags in the back seat. I fired up the engine and headed for the gate as fast as I could without running anyone over.

My mind was already racing ahead, trying to picture the location I'd been given. Shah Shahid, southeast towards the Jalalabad highway and near the Ministry of Defense. I knew the area—upscale neighborhood, wide streets, some of the nicer houses in Kabul. A lot of government officials lived there. A lot of civilians.

As I cleared the gate and turned onto the main road, I keyed my radio. "Julliett Sierra, departing for Xray."

The acknowledgment crackled back through static. I pushed the accelerator down and wove through the dense morning traffic, my hands tight on the wheel.

"Do either of you know what happened?" I asked, glancing at my passengers in the rearview mirror.

The Navy Corpsman leaned forward between the seats. "According to the report I heard, a small convoy of military advisors got hit by a VBIED on their way to work. Initial reports made it sound pretty bad."

"Son of a bitch."

My heart dropped into my stomach. The timing clicked into place with sickening clarity. About twenty minutes after the explosion had rattled General Washington's window, I'd gotten the call to transport medics. Morning rush hour. Advisors heading to work. The pieces fell together like a puzzle I didn't want to solve.

The streets blurred past. Concrete walls topped with razor wire. Crumbling buildings pockmarked with old bullet holes. Then I saw it—thin wisps of black smoke rising ahead, growing thicker as we got closer. I rolled down my window. The wail of sirens grew louder, converging from multiple directions. The acrid smell of burning rubber and something else—something chemical and wrong—hit my nostrils.

We were almost there.

"Looks like this is as close as I'll be able to get," I said, pulling off to the side of a residential street about a block from the site.

The road ahead was choked with vehicles. MRAPs with their angular, blast-resistant hulls. Fire trucks. Ambulances. Afghan National Army pickups. All of them parked at chaotic angles, blocking the street.

Both medics sprang into action before I'd even shifted into park, grabbing their bags and sprinting toward the smoke. I killed the engine, grabbed my M4 from the passenger seat, and took a long, deep breath. The air tasted like ash.

I opened the door and stepped out into hell.

Nothing—no training, no briefing, no mental preparation—could have prepared me for what I saw.

The quiet, upscale neighborhood had been transformed into a warzone. In the middle of the street ahead of me, sat the smoldering skeleton of what had once been an armored Chevrolet Suburban. The reinforced steel frame. The engine block. The melted remnants of bullet-proof windows. That's all that remained. The VBIED had reduced a vehicle designed to withstand small-arms fire and IEDs into a pile of twisted, blackened metal.

Everyone inside—vaporized.

Thirty feet behind the Suburban's remains, a crater the size of a hot tub had been punched into the asphalt. Jagged chunks of road radiated outward like frozen waves. That's where the suicide bomber had detonated—timing it perfectly to drive between the lead and rear vehicles of the convoy.

But it wasn't just the convoy that had been destroyed.

An entire block of what had once been beautiful two-story apartments, family shops, and houses had been decimated. Walls were blown out, exposing the intimate details of people's lives—a kitchen table set for breakfast, curtains fluttering through frames that no longer held glass, a child's tricycle lying in what used to be a living room. Every window within a hundred yards had shattered. The blast wave had reached inside these homes and torn them apart.

The bomb had gone off during breakfast time. Children had been eating naan and eggs at their kitchen tables, getting ready for school. Mothers had been braiding their daughters' hair. Fathers had been sipping tea and preparing for work. And then, in a single instant, their worlds had exploded.

The scene was absolute chaos. Ambulances screamed in and out, their drivers navigating around debris and bodies—some moving, some not. Afghan emergency workers in mismatched uniforms shouted in Dari, trying to coordinate the evacuation of the wounded. I watched as two of them carried a small boy, maybe seven years old, onto a stretcher. Blood streamed from his ears and nose. His school clothes—a white shirt and dark pants that his mother had probably ironed that morning—were shredded and soaked red. His eyes were open but unfocused, staring at nothing.

A woman in a burqa stumbled past me, her robes soaked with blood—hers or someone else's, I couldn't tell. She was wailing, calling out a name over and over. "Jamila! Jamila!" The sound cut through every other noise on that street.

The medics I'd brought were already on their knees beside a teenage girl whose left leg had been partially severed below the knee. The French doctor worked with his hands, his fingers moving quickly while he shouted orders I couldn't understand. The Navy Corpsman was applying a tourniquet, his hands steady despite the girl's screaming. Blood pooled beneath them, spreading across the concrete.

More ambulances arrived. More stretchers. More screaming. The wounded being loaded into vehicles were mostly women and children. Civilians. Non-combatants. People whose only crime was living in a neighborhood where a terrorist decided to make a statement.

I saw one mother clutching an infant who wasn't moving, would never move again, though she kept trying to wake the baby up. She rocked back and forth on her knees, pleading with God or fate or anyone who might listen, her voice hoarse from screaming. An old man sat on the curb twenty feet away, holding his head, blood pouring between his gnarled fingers. He stared at nothing, his eyes empty.

I walked toward a cluster of American ISAF soldiers who were standing behind what was left of the trailing vehicle, surveying the damage and talking into radios. Catching sight of my rank insignia, an Army Sergeant spun to face me.

"Sir, how can we help you?"

"Hi, Sergeant. I'm from HQ. I'm just trying to find out what the hell happened here. Who got hit?"

He pulled a small notebook from his cargo pocket, flipping it open. "We're still piecing together the details, sir, but from what our teams have gathered so far from eyewitnesses, a white Toyota Corolla accelerated between the two vehicles right here"—he pointed to the crater—"and detonated. We think the lead vehicle was the security detail. The rear vehicle probably carried the ISAF advisors."

My mind raced. I worked with a lot of advisors during my weekly basing coordination meetings. Their security details—often Army soldiers—acted as guardian angels when we met with the Afghans at different bases across the region.

"We still need confirmation," the Sergeant continued, "but we think the security detail was two Army soldiers from the Guam National Guard. A Sergeant and a Specialist. Two young guys. They deployed here last month with the 294th."

I knew the unit. They had arrived in April, about the same time I did. The 294th had a strong presence at almost every base in Afghanistan. They were squared away—professional, disciplined, always calm under pressure. They conducted numerous security operations across Afghanistan, including personal security details and guardian angel escort missions. I always felt safer knowing they were around.

Today, two of them had performed the ultimate sacrifice.

"And the advisors?" I asked.

"Not sure yet sir. Likely ISAF contractors from the Combined Security Transition Command-Afghanistan at Camp Eggers. We're working on positive IDs, but..." He trailed off, glancing at the smoking wreckage.

There was nothing to identify.

Camp Eggers. It was located in the Green Zone next to HQ ISAF, and I'd walked over there just last week for a coordination meeting. Had probably shaken hands with some of these guys. Had probably sat across a table from them, drinking bad coffee and complaining about the bureaucracy.

My heart felt like a stone in my chest.

"Thank you for the information, Sergeant. I'll let you get back to it. Take care."

I walked back toward the blast site, my boots crunching over broken glass and chunks of concrete. The parked cars lining both sides of the street had been shredded. Every single vehicle was pockmarked with thousands of tiny puncture wounds—shrapnel from the explosion had torn through sheet metal like tissue paper.

But it wasn't the damage to the cars that made my stomach lurch.

On the side panel of a smoke-singed sedan, I saw what looked like cooked ground beef that had been thrown against the metal. Chunks of it. Smeared and stuck. Dark red and black. On the wall of a house ten feet away, more of the same. Clinging to the concrete.

It took my brain a few seconds to catch up with what my eyes were seeing.

These weren't pieces of car upholstery or debris from the explosion.

This was human tissue. Cooked. Charred. Seared.

The VBIED's heat had been so intense—thousands of degrees in the first milliseconds of detonation—that flesh had literally been flash-cooked and thrown against any surface within the blast radius. The suicide bomber. The soldiers in the security vehicle. Maybe even some of the civilians caught too close. All of them reduced to meat and scattered across the street like debris.

I turned away and dry heaved, bending at the waist, my hands on my knees. Nothing came up. My stomach was empty. But my body tried anyway, convulsing, trying to reject what I'd just seen.

And then the smell hit me. Really hit me.

Burning rubber from tires that had melted in the heat. Thick smoke that hung in the morning air like fog. Scorched metal. Chemical accelerant from the explosives.

And underneath it all, threading through everything else, was something sweet and rotten and fundamentally wrong. Like meat left on a grill too long. Like every nightmare smell you've ever imagined concentrated into a single breath.

I'd never forget that smell. It embedded itself in my nostrils, my clothes, my hair, my memory.

I straightened up and forced myself to keep moving, careful not to step on anything that might have once been part of someone. Afghan National Army and Police members were setting up a perimeter with police tape, trying to keep the growing crowd of civilians back.

That's when I saw them.

About fifty feet from the crater, in front of a house with a section of the exterior wall blown out, a man and woman knelt on the ground. Between them lay a small form covered by a white sheet, already soaking through with bright red blood. The stain spread slowly, blooming like a flower.

The father had his hands on his knees, his head bowed low, his shoulders shaking with silent sobs. He wore a traditional shalwar kameez that was torn at the shoulder. His feet were bare. He must have run outside without his shoes when he heard the explosion.

The mother was bent over the body, her wails cutting through every other sound on that street like a knife. She was calling a name—her daughter's name—touching the sheet with trembling hands, pulling back as if the touch burned her, then reaching out again. Over and over. A cycle of grief too enormous to contain.

"Laila. Laila, my Laila."

I stood there, frozen, watching them. Useless. Helpless. My throat tightened until I could barely breathe. My eyes burned.

I thought about my son back home. His laugh—that pure, uninhibited belly laugh that kids have before the world teaches them to be self-conscious. The way he'd run to me when I came through the door after work, his little legs pumping, his arms outstretched. How he'd wrap his arms around my neck and squeeze with all his strength.

I tried to imagine losing him. Tried to imagine being that father, kneeling in the street over my child's broken body, knowing that just an hour ago she'd been alive—laughing, eating breakfast, asking to play with his Lego's a little longer before leaving for school.

My mind wouldn't let me go there. Couldn't go there. The pain was too enormous to even approach. Too fundamental. Too wrong.

But that father didn't have a choice. He was living it right now. Would be living it for the rest of his life. Every morning, he'd wake up and remember. Every breakfast, every school day, every holiday. The empty chair at the table. The silent bedroom.

I don't know how long I stood there. Could have been thirty seconds. Could have been ten minutes. Time moved strangely in that space, stretching and compressing like something elastic.

Eventually, someone shouted nearby. They needed help securing the perimeter. I forced myself to move. I went where they pointed. I did what they asked. I tried not to look at the small shoes scattered in the street like abandoned toys—a brown sandal, a black sneaker, a tiny leather shoe meant for a toddler.

But I saw them anyway. All of it. Every detail burned into my memory like a brand.

By the time I decided to return to base, the sun had climbed higher, and the smoke had finally started to thin. The worst of the wounded had been evacuated. The dead were being covered with blankets and sheets. Someone would come later to collect them, to try to identify them, to return them to their families for burial.

I walked back to my Land Cruiser alone—the medics had left in an ambulance with critical patients. My hands were shaking as I opened the door. I gripped the steering wheel hard, trying to stop the tremors. It didn't work.

The final count, relayed over the radio as I sat there: All six ISAF personnel dead. Killed instantly. They probably never knew what hit them. One second they were alive, thinking about their meetings, maybe making plans for lunch. The next second, nothing.

Fifteen civilians confirmed dead.[1] Most of them children. Over fifty more seriously injured—burns covering half their bodies, shrapnel embedded in organs, crushed limbs, traumatic brain injuries. Injuries that would follow them for the rest of their lives.

The bomb didn't discriminate. Didn't care about innocence or guilt. It just destroyed everything in its radius.

---

[1] "Kabul Suicide Bombing of NATO Convoy Kills 15," *ABC News*, 16 May 2013, https://www.abc.net.au/news/2013-05-17/kabul-suicide-bombing-of-nato-convoy-kills-15/4695370.

I keyed my radio. "Julliett Sierra, returning to base."

My voice sounded flat, mechanical, like it belonged to someone else.

As I drove back through Kabul's streets, I couldn't stop seeing that mother bent over her daughter. Couldn't stop hearing her wails. Couldn't stop replaying the morning's events in my head.

My hands wouldn't stop shaking on the wheel.

This was the reality. Every single time I left the wire, this was the reality waiting outside the gate. The insurgents didn't need sophisticated tactics or advanced training. They just needed a car, some explosives, and someone willing to die.

I thought about all the times I'd driven these same streets, heading to meetings at bases scattered across Kabul. All the times I'd convinced myself that the odds were in my favor. That it wouldn't be me. That today wasn't the day.

But probability doesn't mean shit when you're the one in the blast radius.

I pulled through the base gate, showed my ID to the guard, waited for them to sweep my vehicle for bombs, and headed for the motor pool. I sat there for a long moment with the engine off, my hands still gripping the wheel, staring at the concrete walls ahead of me.

Behind me, outside the wire, people were still dying. Families were identifying bodies. Hospitals were running out of blood and supplies. But inside the wire, soldiers were walking to chow, complaining about the gym equipment, checking their email, living their day.

Two different worlds separated by a wall.

I looked down at my uniform. Black soot was smeared across my boots and pants. Blood—someone else's blood, none of it mine—was spattered across my left sleeve. I'd need to change before heading back to the office. Before I could sit at my desk and answer emails and coordinate basing operations as if I hadn't just spent the morning standing in a street full of dead children.

I climbed out of the Land Cruiser and walked toward my Conex, each step feeling heavier than the last. The smell clung to me—burning rubber, cooked flesh, smoke, blood. It was in my nose, my hair, my skin. I'd shower. Change clothes. Wash my face.

But I knew, even then, that I'd never really wash it off.

Yet as I walked, something unexpected began to crystallize in my chest. Not healing—that would take years, if it ever came at all. But something harder. A kind of resolve forged in the crucible of witnessing the worst humanity could do to itself.

I'd seen evil today. Pure, calculated evil designed to maximize suffering. But I'd also seen something else. The medics who'd sprinted toward danger without hesitation. The Afghan emergency workers risking their lives to save strangers. The father and mother who, even in their unbearable grief, had stayed with their daughter, refusing to let her be alone.

Maybe that was what survival looked like in a place like this. Not the absence of brokenness, but the stubborn refusal to let brokenness have the final word. Not the elimination of pain, but the determination to find meaning in spite of it.

I thought about Jordan, back home in Colorado, fighting his own battles of losing his mom and sister. About the promise I'd made to come home to him. About the man I wanted to be when I walked through that door— not unmarked by this place, but not destroyed by it either.

I was learning something about resilience that they don't teach in leadership courses. It wasn't about being strong enough to avoid breaking. It was about being stubborn enough to keep functioning even after you've shattered into pieces.

The shower was cold, but it felt good. The clean uniform felt better. And when I walked back into the Basing Office an hour later, I carried myself differently. Not because I felt invincible—quite the opposite. But because I'd looked into the abyss and decided I wasn't going to let it swallow me whole.

"How'd it go out there?" Bill asked quietly, looking up from his paperwork.

"Bad," I said simply. "But the medics did an amazing job saving lives today. Everyone did their best to lend a hand."

He nodded, understanding in his weathered face. "That's all any of us can do."

I sat down at my desk—Landry's desk, my desk now—and pulled up the day's schedule. There were meetings to coordinate, reports to file, bases to transfer. The machinery of war didn't stop for individual tragedies, no matter how devastating.

But I was different now. Tempered by fire. Baptized by blood and smoke and the sound of a mother calling her daughter's name.

I was Chief of Basing for NATO in Afghanistan. I had work to do. The war would continue. The bombs would keep going off. More names would be read at memorial services. I would be here, carrying my piece of the load—not because I was strong enough, but because I was broken enough to understand what was at stake, and stubborn enough to keep fighting anyway.

The resolve lasted until I reached my quarters.

Then the adrenaline drained out all at once, leaving nothing but exhaustion and the images I'd been holding at bay all afternoon. I sat on the edge of my cot and stared at the wall. The wall offered nothing back.

Sleep wouldn't come. Every time I closed my eyes, I was back in the street. The mother bent over her daughter's body. Cooked flesh smeared across walls and cars. Small shoes scattered in the dust.

I sat up, hands shaking, and reached for my phone. I needed to hear Jordan's voice more than I needed sleep.

I slipped outside so I wouldn't wake anyone, easing the door shut behind me. The night air hit me—diesel, dust, the faint chemical bite of a city that had been burning one way or another for forty years. Above it all, the stars over Kabul were indifferent and brilliant, the kind of sky that doesn't care what happens beneath it.

The time zone math took a moment through the fog in my head. 2400 hours in Kabul meant 1430 in Colorado. School should be letting out. Maybe my mom was already in the pickup line.

My hands trembled as I dialed.

The connection clicked and hissed through satellite delay. Then, finally, it rang.

"Hello?" My mother's voice—familiar and impossibly far away.

"Mom. It's me." My voice sounded raw.

"John! Oh, it's so good to hear from you. Are you okay? You sound—"

"I'm fine." The lie tasted like ash. "Is Jordan there? I just… I needed to hear his voice."

A pause. Long enough for her to understand what I wasn't saying.

"I just arrived at his school. The parent pickup line is long today. Can you call back in ten minutes?"

Ten minutes.

"Yeah. I can do that."

"John." Her voice softened into that tone mothers use when they know their children are hurting. "What happened?"

"Just a bad day, Mom. A really bad day. I'll call back in ten. Love you."

I ended the call before she could press further, before the concern in her voice cracked the control I was barely holding together.

I stood there counting seconds, trying not to see the little girl under the white sheet. Trying not to hear her mother's wails. Trying not to think about how many parents in Kabul were tucking children into empty beds tonight.

Jordan was safe. In Colorado. Getting out of school. Probably talking about recess.

Jordan was alive.

The phone rang in my hand, startling me so badly I nearly dropped it.

"Dad!" His voice burst through the speaker with the unfiltered energy only seven-year-olds possess. "Grandma said you were going to call back and I've been waiting forever—well, not forever but like a while—and guess what happened at school today?"

The sound of his voice hit me like a physical blow. I closed my eyes and pressed my free hand over my mouth to keep whatever was rising in my chest from breaking loose.

He was alive. Happy. Untouched.

"What happened, buddy?" My voice cracked, and I cleared my throat, hoping he'd blame the connection. "Tell me about your day."

"So we had this science thing, right? And Mr. Peterson—he's so cool, Dad, you'd like him—he brought in this volcano. Not a real-real one, but one he built, and we made it explode with vinegar and baking soda and it went everywhere! Megan got it in her hair and she got sort of mad but Mr. Peterson said that's what science is supposed to do—make a mess sometimes—and—"

He kept going, tumbling from volcanoes to pizza at lunch to kickball at recess, where he almost kicked a home run but didn't quite, to the new kid from California who had never seen snow.

I listened to every word, clutching the phone until my knuckles went white, letting the normalcy wash over me like a baptism. This was real. This innocence. This excitement over classroom volcanoes and kickball games. Not the burning vehicles. Not the scattered shoes. Not the smell that still clung to my uniform.

"That sounds amazing, Jordan," I said when he finally paused for breath. "I'm so proud of you, buddy."

"I wish you could've seen it," he said, the bravado slipping, the little boy underneath showing through. "When you come home, maybe we could make one? You and me?"

"Absolutely. First weekend I'm back, we'll build the biggest volcano Colorado has ever seen. Deal?"

"Deal!" The excitement flickered, then softened. "Dad? When are you coming home? Grandma says before Christmas, but that's… that's a lot of days."

I swallowed the lump in my throat. "I know it feels like forever. But I'll be home before Christmas. Maybe even before Thanksgiving."

"You promise?"

The weight of it settled over me. I'd seen death up close that day. But I was still here. Still breathing. Still able to promise.

"I promise, Jordan."

"Okay." His voice brightened. My mother's voice drifted in the background. Then Jordan again. "Grandma says I have to do homework. Can you call again soon?"

"Soon as I can. I love you more than anything in this world."

"Love you too, Dad."

The line went dead.

I stood in the dark, phone pressed against my chest, tears sliding down my face. No one could see. No one would know. Jordan's day had been volcanoes and kickball. Mine had been charred bodies and screaming mothers. His biggest problem was homework. Mine was staying alive long enough to keep a promise.

The contrast felt obscene—the impossible gap between his world and the one outside the wire.

I saw the Afghan girl again beneath the white sheet. Someone's daughter. Someone's entire world. That morning she had probably been excited about something small and ordinary. Now her mother would never hear her voice again.

But Jordan's voice had given me something stronger than fear.

Tomorrow I would put on my body armor. I would drive Kabul's streets. I would do my job. But that night, standing under those indifferent stars, I let myself grieve—for Laila, for her mother, for every empty bed in that shattered neighborhood.

Sleep came eventually.

And for a few merciful hours, I was home.

Chapter 7: Into the Shadows

May 17, 2013

The morning after the bombing, I woke with the smell of smoke still coiled in my nostrils—clinging like a lie, even though I'd showered twice and changed uniforms. The acrid memory of burning flesh and metal had burrowed into my sinuses like a parasite, reminding me with every breath that what I'd witnessed wasn't a nightmare I could wake from. It was just another Friday in Kabul.

I tried to shake it off as I climbed into my vehicle. Today's a new day, a fresh start, I told myself, the words feeling hollow even as I thought them. It was also my first week as the new Chief of Basing—a promotion that had come gift-wrapped in chaos and blood. I never got to say goodbye to Captain Landry. Military police had whisked him onto base under cover of darkness and spirited him away just as quickly with the few belongings my team had packed, leaving behind only questions and the faint smell of desperation.

Now, starting the engine, I felt more than just the heaviness of my body armor pressing against my shoulders. I felt the accumulated mass of responsibility settling into my bones—the transfer roadmap with its 346 bases, the coordination meetings with Afghan officials who trusted me with their nation's future, the knowledge that any mistake I made could cost lives. My hands gripped the steering wheel a little tighter than necessary as I began the very short drive to the US Embassy Annex for my first meeting with the CIA.

I arrived with fifteen minutes to spare, using the extra time to scan my surroundings with the paranoia that had become second nature. The Annex sat behind its own fortress of walls and barriers—a former landmark transformed into a shadow compound.[2] What had once been the Ariana Hotel, Afghanistan's first Western-style luxury hotel when it opened in 1945, now housed the real war, one fought with intelligence and midnight operations rather than tanks and artillery. The building's grand bones were still visible beneath the fortifications. A ghost of the cosmopolitan Kabul that existed before decades of conflict turned local landmarks into hardened targets. As I walked up to the building, a large burly man in his mid-forties emerged from the entrance. He wore jeans, boots, and a long-sleeve brown denim shirt with the sleeves rolled back—the unofficial uniform of contractors who wanted to blend in but couldn't quite hide the military bearing that years of service had burned into their posture.

"Hi Major, you must be Landry's replacement," he said, his voice carrying the gravel of someone who'd smoked too many cigarettes in too many safe houses. His eyes—pale blue and unnervingly direct—examined me from head to toe with the clinical assessment of a butcher sizing up meat. "I'm Chris, but I go by Ghost."

The nickname fit. There was something insubstantial about him, as if he could fade into shadows at will. His face was weathered leather, creased

---

[2] Gossman, Patricia. "The CIA's Afghan Ghosts': The Ariana Hotel Compound." Just Security, 14 Sept. 2021, www.justsecurity.org/78234/the-cias-afghan-ghosts-the ariana-hotel-compound/.

with lines that came from squinting into foreign suns and watching men die in a dozen different countries.

"Nice to meet you Ghost, I'm John. Starting today I'm taking over all of Landry's duties… to include this one I guess." I tried to sound more confident than I felt, keenly aware that I was stepping into a world I didn't fully understand.

"Damn shame about Landry. Good fella." Ghost's expression didn't change, but something flickered behind his eyes—maybe sympathy, maybe calculation. In his world, people broke all the time. The only question was when and how spectacularly.

For the next few minutes, in what felt like an interrogation disguised as small talk, Ghost asked a series of questions about my background and duties in the Air Force. Where I'd served. What clearances I held. Which Task Force operations I'd supported. Each answer seemed to unlock another question, each one probing deeper, testing not just my credentials but my character. Could I be trusted? Could I keep secrets? Would I crack under pressure like Landry, or was I made of sterner stuff?

Finally, seeming to approve of my experience over the previous decade, he gave me the nod of approval—a slight tilt of his head that somehow conveyed both acceptance and warning.

"Almost meeting time. Follow me," he said, turning on his heel and leading me down the sidewalk into the building.

Ghost led me down a narrow corridor that smelled faintly of coffee and paper, then into a small backroom with walls made of dark wood paneling. The space was stuffy and windowless, the air conditioning struggling against the Afghan heat. It felt like a coffin designed for classified conversations.

Standing inside were two men who couldn't have looked more different. The first was tall and rugged, maybe thirty-five, with a weathered face that came from field operations rather than desk work. His eyes held that thousand-yard stare I'd seen in soldiers who'd spent too long watching bad things happen through night-vision goggles. The second man was smaller, almost bookish, wearing glasses, a polo shirt, and khaki pants that made him look like he'd wandered in from a suburban office park.

"This is Tom," Ghost said, pointing to the larger man with a gesture that suggested long familiarity. "He works Intel." He chuckled as he pointed to the man in glasses, and I caught a glimpse of genuine fondness beneath the professional distance. "And this nerd here is Allen. He's with another organization but works closely with us. He's pretty much a whiz kid when it comes to anything dealing with signals intelligence."

"Nice to meet you both," I said, greeting them with warm handshakes that felt absurdly normal given the circumstances. We were about to discuss things that could get people killed, and we were starting with pleasantries.

"Welcome to the team, Maj Spargur," Tom said, his voice carrying an Oklahoma drawl that softened the edges of his command presence. "Please sit down. Let's get you up to speed on why you're here."

We all settled around a wooden rectangle coffee table that had probably hosted a hundred similar conversations—trading secrets, coordinating operations, connecting the dots that civilians would never see. In front of Tom sat a brown folder. Whatever was about to be discussed, it mattered.

Tom pulled out a sheet of paper I recognized immediately—my own ISAF Theater Basing Roadmap, covered in his handwritten notes and highlighting.

"Let's talk about this first," he said, turning the paper so it faced my direction. The map showed Afghanistan carved into colored sections, each base marked with symbols indicating its status, classification, and projected closure or transfer date. It was my life's work since I'd arrived at HQ, reduced to a single page of strategic importance.

"We follow your ISAF Basing Roadmap very closely," Tom continued, his finger tracing routes between bases like he was planning an operation— which, I realized with a chill, he probably was. "There are, ummm… let's just say very sensitive equipment at many of the strategic and operational bases listed on your Roadmap. Equipment that can't fall into the wrong hands. Equipment that needs to be extracted before those bases change ownership."

He paused, letting that sink in. I thought about all the bases I'd been coordinating transfers for—Forward Operating Bases scattered across

Helmand and Kandahar, Combat Outposts in remote provinces, strategic installations like Bagram and Kandahar. How many of them housed Agency listening posts? Drone control stations? SIGINT collection facilities that intercepted Taliban communications?

"Now that NATO is downsizing their basing footprint," Tom went on, "the CIA needs to be the first to know about any changes to the basing plan. We're using the timelines from your Roadmap to plan our own exfiltration schedules, and if you make any adjustments—even small ones—we need to know ASAP so that we have time to adjust our plans without impacting our mission or, more importantly, without compromising our assets in the field."

Assets. The word hung in the air with all its implications. Not just equipment. People. Afghans who'd worked with the Agency, who'd risked their lives providing intelligence, who'd be marked for death the moment the Taliban discovered their cooperation. Their lives depended on coordinated timing—pulled out too early and you blew their cover; too late and they'd end up in a ditch with a bullet in their skull.

"Politics are getting pretty squirrely now that we're coming close to ending our time here," Tom continued, and I heard the frustration in his voice. "ISAF has been changing policies weekly, hell, sometimes daily. Generals back in Washington making decisions without understanding ground truth. Politicians worried about optics rather than outcomes. With you sitting at the strategic HQ next to Gen Dunford and working with Gen Milley's people at IJC, we need you to be our eyes and ears for any important updates. Make sense? Any questions?"

The enormity of what he was asking crystallized in my mind. I wasn't just managing base closures anymore. I was a key node in an intelligence network that stretched across Afghanistan. Every email I sent, every meeting I coordinated, every timeline I adjusted could trigger a cascade of covert operations I'd never see.

"Sure, you can count on me to closely coordinate any basing updates," I said, keeping my voice steady despite the tension climbing up my spine. "I'll make sure you guys have a chance to review each new edition of the Basing Roadmap before I publish it."

"Perfect, that's what we wanted to hear," said Tom, and I caught the ghost of a smile—approval from a man who didn't give it easily.

"Alright, there's one more matter we need to discuss," Ghost said, his voice taking on a different quality. Tom reached back into his folder, this time pulling out an assortment of photographs and spreading them across the table like a dealer laying out cards in a high-stakes game.

"These are persons of interest," Ghost said, his finger tapping one photograph in particular. "Possible facilitators, weapons smugglers, IED makers, informants who've gone dark. Some work in the Afghan government. Some work construction on your bases. Some drive the trucks that bring your supplies."

And for the next half hour, we delved into a world I'd never experienced before—a world cloaked in shadows and populated by ghosts. My head spun as I struggled to absorb the details of CIA counterterrorism operations in Afghanistan and the small but crucial role I would play. Each photograph came with a story: This man arranged funding for Taliban operations from donors in Pakistan. That one built IEDs in his house and taught others the craft. This government official took bribes to look the other way when insurgents moved through checkpoints.

Ghost spoke with the clinical detachment of a surgeon describing tumors, but the implications were anything but clinical. These men moved through the same streets I drove on. Ate at the same restaurants. Some probably worked on the bases I was responsible for closing and transferring. The enemy wasn't just "out there" in the mountains and villages—the enemy was woven into the fabric of everything we were trying to accomplish.

Tom explained targeting packages and kill chains, how intelligence moved from collection to analysis to action. Allen, the quiet analyst, walked me through the electronic signatures they tracked—cell phones, internet usage, financial transactions. Everything left a trail if you knew where to look, and the Agency had become very good at looking.

"This is where you come in," Ghost said, spreading out a map of Kabul marked with colored pins. "Several of these individuals operate from bases on your closure list. We need advance notice before you transfer those facilities to the Afghans so we can adjust our operations accordingly. I cannot overstate the importance of keeping us informed—not just about your timeline, but about the planning meetings with Afghan officials leading up to each transfer."

I stared at the map, trying to memorize the pin locations, understanding that my coordination meetings with Afghan officials now carried an additional layer of complexity. Every transfer I negotiated, every timeline I adjusted, had to be filtered through this new requirement.

Sensing I was oversaturated with information—my brain reaching that point where new data just slides off the surface rather than sinking in—Tom scooped up the photos and papers from the table and shuffled them back into the folder with practiced efficiency.

"That's enough fun for one day," he said, and I detected dark humor in his voice. This was his daily reality—tracking threats, connecting dots, coordinating operations that would never make the news. "We appreciate your willingness to support, Maj Spargur. Let's plan on meeting here at the same time next week. If something comes up before then, you know how to reach Ghost."

And with that, I was ushered out the same way I came in, Ghost's hand on my shoulder guiding me through the corridor and back into the harsh Afghan sunlight.

The brightness hit my face like a slap after the dim conference room, making me squint as my eyes adjusted. I stood there for a moment, disoriented, as if I'd emerged from an underground bunker to discover the world had changed while I was below ground. In a sense, it had. My understanding of my role, my mission, my place in this war—all of it had been fundamentally altered in the span of half an hour.

I carried new secrets now, far different from the ones I'd held about nuclear missiles and satellite operations during previous assignments. Those had been abstract threats—devastating but distant, ICBMs that would hopefully never launch, satellites tracking objects in the cold vacuum of space. These secrets were immediate and personal. Names. Faces. Targets. Operations that would happen in the next days and weeks, potentially saving or ending lives based on decisions I helped facilitate.

The significance of it pressed down on my shoulders more heavily than any body armor. I was holding threads now that could unravel everything—not just wearing a uniform and following orders. The Basing Roadmap that could make or break CIA operations. The trust of men who lived in shadows and whose failures meant body bags. The responsibility of being Gen Dunford's eyes on the ground while simultaneously serving as the Agency's early warning system.

I climbed back into my vehicle and sat there for a moment, gripping the steering wheel and staring at the concrete blast barriers ahead without really seeing them. How the hell did a kid from small-town America end up here? The question had no good answer. A series of choices, some mine and some made for me, had led to this moment—sitting in a Toyota Land Cruiser in Kabul, Afghanistan, carrying secrets that would get me killed if the wrong people discovered what I knew.

I thought about Jordan back home, about the promise I'd made to return. That promise felt more complicated now. I wasn't just trying to survive rockets and IEDs anymore. I was neck-deep in a shadow war where the boundaries between friend and enemy blurred, where yesterday's ally could be tomorrow's informant, where trust was a luxury and paranoia was survival.

Starting the engine, I headed back to headquarters, carrying a burden no one around me would ever know about. The other soldiers I passed—walking to the DFAC, heading to the gym, standing in clusters talking and laughing—they lived in a different war than I did now. Their battles were visible: firefights and patrols, convoys and checkpoints, the daily grind of military operations in a combat zone.

My war had just become invisible. Fought in conference rooms and classified briefings, measured in intelligence reports and surveillance photos, won or lost based on timelines and coordination that would never earn medals or citations. Catastrophic events leave scars, but the secrets surrounding them can be even more damaging to those forced to carry them. The isolation of knowing what others don't, of holding truths that cannot be shared, exacts a heavy toll. Landry had been weathered by the burden—hollowed out in subtle ways that manifested in his guarded demeanor and the shadows behind his eyes.

As I drove through the gates of HQ ISAF, past the guards and barriers and weapons that created the illusion of security, I made a conscious decision: I would not break. Not like the others who'd come before me and discovered that the weight of classified knowledge was heavier than any physical burden.

I would do my job. Coordinate my meetings. Maintain my roadmap. And I would keep the secrets that needed keeping, because lives—both American and Afghan—depended on my discretion and competence.

But I would also keep one piece of myself separate, protected, walled off from the machinery of intelligence operations and covert wars. I would remember why I was here: to survive, to get home, to keep the promise I'd made to a seven-year-old boy who needed his father.

Everything else was just noise.

I parked in the motor pool and killed the engine, then sat in the sudden silence, listening to the tick of cooling metal and the distant sound of helicopters. Around me, the base hummed with controlled activity—the normal rhythm of military life in a war zone.

But I wasn't the same person who'd driven away an hour ago. I'd crossed another threshold, this one invisible but just as real as the one I'd crossed when my C-130 first landed in Kabul.

I was in the shadows now. And like Ghost had said with that barely perceptible nod, the only question was whether I could handle living there.

Time would tell.

I grabbed my rifle, stepped out of the vehicle, and walked toward my office to finish my first full day as Chief of Basing.

The sun beat down. The dust swirled. Afghanistan kept turning, indifferent to the small dramas of the Americans who thought they could control it.

And somewhere in the city beyond the wire, the men in those photographs were going about their daily routines—eating breakfast, making phone calls, building bombs, planning attacks—completely unaware that their faces were pinned to a map in a windowless room where men like Ghost and Tom tracked their movements and waited for the right moment to act.

The game continued. I was just another player now, trying not to get checkmated.

Chapter 8: Fighting Season

June 10, 2013

Spring hit Afghanistan like a door being kicked open. The snow that had blanketed the peaks of the Hindu Kush through winter began its retreat, exposing the rocky vertebrae of mountains that had watched empires rise and fall for millennia. With the melting snow came something else—the Taliban's annual spring offensive, code-named "Khalid bin Waleed" after the Islamic general known as the "Drawn Sword of God."

The offensive started early this year—too early. Usually, the Taliban waited until May when the mountain passes were fully clear and the heat made travel easier. But 2013 was different. Intelligence reports showed a significant increase in attacks compared to the previous year. They were testing the Afghan security forces, probing for weakness, and finding plenty of it. Now, with the rat trails from Pakistan running full tilt, hundreds of fighters were streaming across the border every day like water finding cracks in a dam.

Fighting season wasn't coming. It was already here, and it had teeth.

I rolled into the office that morning expecting the usual routine—coffee that tasted like burnt motor oil, planning meetings that accomplished little, the endless shuffle of paperwork that kept the bureaucratic machinery grinding forward. Instead, Rod looked up from his computer with that expression I'd learned to dread over the past months: the tightening around his eyes, the slight downward pull of his mouth, the body language that said our day just got significantly worse.

"Looks like we'll have to cancel our trip to NKAIA today," he said, his normally cheerful Newfoundland accent carrying an edge of frustration.

"Why's that?" I asked, though my gut was already churning with the answer. In Afghanistan, cancelled meetings meant only one thing: somebody was trying to kill somebody, and it was too dangerous to drive through the crossfire.

Rod leaned back in his chair, the springs creaking in protest. "The ISAF Intel report just came out. Apparently, the Taliban are trying to take down ISAF Joint Command at NKAIA. They started attacking at 0430 this morning. Shut down both the civilian and military sides of the airport. Snipers, heavy machine guns, and rocket propelled grenades are keeping the ISAF and ANA forces at bay so far."

He paused, scrolling through the report on his screen, his face illuminated by the harsh glow of the monitor. "Oh, and the Taliban executed two boys in Kandahar that they accused of spying, and blew up a military convoy in Ghazni, killing a Polish troop."

The last detail hit different. "Damn. Does Dombroski know yet?" I was thinking of the quiet Polish officer who worked in our section, who kept a picture of his two kids taped to the edge of his computer monitor—a boy and a girl with matching gap-toothed smiles, frozen in some happier moment before deployment.

"Yeah, he said he didn't know the guy that got killed, but he had a friend in that convoy who got beat up pretty bad from the blast. He's outside taking a smoke, trying to calm his nerves."

I rubbed my eyes, feeling the stress of another day in Kabul settling onto my shoulders like wet concrete. "C'mon Rod, don't you have any good news?"

"Yeah."

"Oh good, what's that?"

"We don't have to go outside the wire today," he said, his face breaking into a dark grin as he shrugged his shoulders.

"True, but it's just postponing the inevitable." I tried to find comfort in the delay, knowing that each day the roads were becoming more dangerous, that each convoy was rolling dice with worse odds. The Taliban were learning our patterns, studying our routes, placing their IEDs with the patience of fishermen setting traps. "I guess I'll take that as a win for today."

The office settled back into its rhythm—keyboards clicking, phones ringing, the low murmur of conversations in a half-dozen languages as coalition partners coordinated their small pieces of a massive puzzle. But the tension was palpable, humming just beneath the surface like high-voltage current running through inadequately insulated wire. Everyone felt it. Spring offensive. Those two words changed everything.

Just then the door opened and General Washington stepped through the doorway, his angular frame filling the space with a presence that came from years of training and combat deployments. I stood up from my desk automatically, muscle memory of military courtesy overriding my exhaustion. This was strange. Washington usually summoned us into his office if he wanted something, preferring to maintain the physical and emotional distance that came with command. He rarely crossed the hallway into our workspace—it was like he wanted to keep us at arm's length, interacting with us only when absolutely necessary.

"Have you guys heard about the attack on IJC?" His voice was flat, professional, but I caught something underneath—maybe concern, maybe just the accumulated stress of managing operations in a combat zone for months on end.

"Yes, sir. We're canceling the basing meeting we had scheduled there for this afternoon." I tried to sound disappointed rather than relieved, though truthfully the cancelled meeting felt like a stay of execution.

"Probably best. Let's hope they get this wrapped up soon." He shifted his stance, and for the first time I noticed something different in his voice— a lighter tone, maybe relief, maybe anticipation. The emotional flatness that usually characterized his speech had cracked slightly, revealing something underneath. "My replacement flies in tomorrow afternoon to NKAIA. Major Spargur, I'll need you to give me a ride to pick him up. We'll need to leave at 1600 hours."

My stomach tightened involuntarily, a physical reaction I'd learned to trust over months of deployment. Another trip outside the wire. Another roll of the dice. Another opportunity to end up as a name read at one of those weekly memorial ceremonies, reduced to a photograph and a folded flag.

"Roger sir, I'll swing by your office ten minutes prior."

"Thanks." He scanned the office once more, his eyes moving over each of us without really seeing us, cataloging bodies and equipment with the detached assessment of a commander taking inventory. Then he turned and walked back across the hallway, the door closing behind him with a soft click that felt oddly final.

"Lucky you, John," Bill said, standing up from his chair to stretch, his weathered face breaking into a smirk that didn't quite reach his eyes. The former Green Beret had seen enough combat to know that "lucky" was relative in Afghanistan—you were lucky until you weren't, and the transition happened faster than you could blink. "How many trips will that be this month?"

"Too many," I said, shaking my head. The number had stopped mattering weeks ago. Each trip felt like the first and last simultaneously, a fresh exercise in controlled fear that never quite became routine no matter how many times you did it.

"What do you think General Washington's replacement will be like?" Rod asked, mercifully changing the subject with the tact of someone who understood when to pivot away from dangerous territory.

"I don't know, but I sure hope he's got more of a personality than Washington," Bill said, taking a long sip of his coffee—the bitter brew we all drank not for enjoyment but for the caffeine that kept us functional through fourteen-hour days. "The man's about as warm as a regulations manual written by lawyers."

Rod chuckled, the sound genuine despite the circumstances. "Seriously. I don't think I've ever seen him crack a smile. You'd think he was born in a regulation handbook or something, just emerged fully formed with perfect military bearing and zero sense of humor."

Bill laughed, and I found myself smiling despite everything. This was survival too—not just the physical kind, but the emotional resilience that came from finding humor in small things, from building camaraderie in a place designed to break you down. "You know, at this point, I wouldn't be surprised if he was."

I sighed, letting myself sink into the banter like a warm bath after a cold day. It was easier than thinking about tomorrow's trip, about the route to NKAIA, about all the places where an IED could be buried or a suicide bomber could be waiting. "But you know what? As much as we struggle to get him to loosen up, at least he doesn't micromanage us. Could be worse. Could be one of those generals who needs to approve every email and second-guess every decision."

"True, true," Rod said, nodding agreement. "Besides, I think I saw him crack a partial smile when he mentioned his replacement. You know he's got to be excited about finally going home."

"Just hoping he doesn't write us up for being too happy next time he walks through here," Bill grinned, and we all laughed—the dark humor of people who'd learned that sometimes laughter was the only appropriate response to absurdity.

I stood up from my desk, needing to move, needing to do something other than sit and think about tomorrow. Motion was better than stillness.

Action was better than contemplation. "Alright fellas, I'm going to go check on Dombroski and then head over to the HQ building to pick up some documents."

The door closed behind me, muffling the office sounds. Outside, the Afghan sun was already climbing toward its zenith, promising another day where the temperature would hit triple digits and the body armor would feel like a weighted vest in a sauna. A hot breeze hit me the moment I stepped into the open air—that suffocating Kabul heat that made your uniform stick to your skin and turned the dust in the air into something you could taste, gritty and mineral, coating your tongue and throat.

I found Dombroski around the corner of the building, leaning against the wall. He was smoking a cigarette, the smoke rising in lazy spirals into the still air. His face looked drawn, older somehow than it had the day before, as if grief had aged him overnight.

"Hey man," I said quietly, stopping a few feet away to give him space. In Afghanistan, you learned to respect the invisible boundaries people erected around their pain.

Dombroski looked up, his eyes red-rimmed—whether from smoke or tears, I couldn't tell. Probably both. He took a long drag from his cigarette and nodded in acknowledgment. "Major." The word came out flat, exhausted.

He reached into his pocket and pulled out his pack of cigarettes, extending it toward me with a gesture I'd seen him make a dozen times before to various people around the base. Always offering, always generous with what little comfort he could provide. I'd always declined before—a polite "no thanks, I don't smoke" that had become a small ritual between us.

Today I took one.

Dombroski didn't say anything, but something shifted in his expression—recognition, maybe, or appreciation for the small act of solidarity. He flicked his lighter and held the flame steady while I leaned in, the unfiltered cigarette harsh against my throat as I took that first drag. Besides an occasional cigar at celebrations back home, I didn't smoke. But right now, it felt like the right thing to do—standing here with him, sharing this small ritual of stress and nicotine, two men trying to process the unprocessable.

"Just John out here," I said, leaning against the wall beside him. The concrete was already hot from the morning sun, radiating heat through my uniform. "Rod told me about your friend. How's he doing?"

Dombroski was quiet for a moment, staring out at the desolate expanse between buildings where heat waves made the air shimmer and dance. When he finally spoke, his accent was thicker than usual—the way it always got when he was stressed, his careful English breaking down to reveal the Polish underneath.

"Marek. His name is Marek." He said the name deliberately, as if speaking it aloud made his friend more real, more than just another casualty statistic. "We served together in Iraq, back in '08. Ramadi. Some of the heaviest urban fighting you can imagine." He flicked ash from his cigarette with a practiced motion. "They said he lost his right leg below the knee. Shrapnel tore up his abdomen pretty badly, punctured his intestines, and nicked his spleen. They got him stabilized and medevacked to Bagram, but..." He trailed off, shaking his head.

"Christ. I'm sorry." The words felt hollow, inadequate, the verbal equivalent of putting a band-aid on a sucking chest wound. What do you say when someone's friend gets their leg blown off? When their guts get rearranged by shrapnel traveling at supersonic speeds? "Is there anything I can do?"

"Unless you can turn back time or get me on a flight to Bagram, no." He wasn't being hostile—just stating a fact with the blunt honesty that combat veterans developed. Politeness was a luxury, and we'd all learned to dispense with luxuries. "I can't even call him. Even if I could get through the phone system, what would I say? 'Sorry you lost your leg on this shit deployment?' What good would that do?"

I didn't have an answer for that. We stood there in silence, the distant sound of helicopters thrumming somewhere over the city like the heartbeat of the war itself. Blackhawks and Chinooks, moving men and supplies, evacuating wounded, delivering death. The soundtrack of our lives here.

"It's getting worse, you know," Dombroski said after a while, breaking the silence with words that carried the heaviness of accumulated observation. "The attacks. Not just here in Kabul—everywhere. I got an email yesterday from a buddy stationed at a COP in Logar Province. He said they're getting hit almost daily now. Mortars every morning at dawn, small arms fire during patrol changes, the occasional complex attack with suicide bombers and assault teams. They can't even take a shit without wondering if a rocket's going to land on the latrine."

"The spring offensive," I said, the words tasting bitter in my mouth. Operation Khalid bin Waleed—the Taliban's annual declaration of increased violence, their promise to step up attacks as weather improved and mountain passes opened.

"It's more than that." Dombroski took another drag, his hands steady despite everything—the hands of a professional soldier who'd learned to function through crisis. "The Taliban know we're leaving. Not tomorrow, not next month, but we're leaving. Everyone knows it. You can see it in how the Afghan National Army look at us now—they're already calculating what comes next, who they'll need to make deals with to survive. And the Afghan National Police?" He laughed, but there was no humor in it, just bitter recognition of an unpleasant truth. "Half of them are just waiting to see which way the wind blows before they commit. The other half are Taliban sympathizers or too stoned on hash to care. Marek's convoy was supposed to have ANP support—an Afghan National Police escort vehicle. You know where they were when the IED went off?"

I could guess, but I let him tell it.

"Still at the checkpoint, drinking chai and watching it happen. Just sat there while Marek's vehicle got blown twenty feet in the air. Didn't fire a single shot, didn't call it in, didn't do shit. Just watched."

The bitterness in his voice was earned, justified by months of watching our supposed partners fail to show up when it mattered. We'd all seen it— the disconnect between what we were supposed to be accomplishing and what was actually happening on the ground. The mission was nation-building, training Afghan forces to stand on their own, creating a sustainable security structure that could survive our withdrawal. But you can't build a nation when half the people you're training are planning to switch sides the minute you leave, and the other half are too high or too corrupt to function.

"My unit back home—the ones still in Poland—they think this is a peacekeeping mission," Dombroski continued, crushing his cigarette under his boot with more force than necessary. "They see the press releases, the photos of us handing out school supplies and building wells. Winning hearts and minds and all that bullshit. They don't hear about convoys getting hit in Ghazni. They don't hear about those two boys the Taliban executed in Kandahar yesterday. Fourteen and sixteen years old. Shot them in the head and left their bodies in the street as a warning to anyone who cooperates with us."

"I heard about that," I said quietly, the image forming unbidden in my mind—two children executed for the crime of talking to Americans, their blood soaking into Kandahar sand as a lesson to their neighbors.

"Everyone heard about it. That's the point." He pulled out another cigarette, lit it with hands that were finally starting to tremble slightly—the adrenaline wearing off, leaving behind the crash. "They want us to know that nowhere is safe. Not Kabul, not the FOBs, not the fucking capital where we're supposed to have control. The attack on IJC this morning? That's right here, where we've got layers of security and checkpoints and all our fancy technology. They've got snipers pinning down coalition and Afghan forces at our main logistics hub. And we can't even leave the wire to go to a meeting because it's too dangerous."

He was right, and we both knew it. The illusion of progress we'd been maintaining—the reports we filed, the metrics we tracked, the PowerPoint slides showing incremental gains—it was all starting to feel like theater. Like we were actors in a play where everyone knew the ending, but we had to keep performing anyway because nobody wanted to admit the script was garbage.

"How long have you got left on your deployment?" I asked, though I suspected he knew down to the day, maybe the hour.

"Four months. 120 days." He said it with the precision of someone who was definitely counting, probably had a calendar somewhere with X's marking off each survived day. "Marek had three weeks left. Three fucking weeks, and now he's missing a leg and probably wondering if his life back home will ever be the same."

The injustice of it hung in the air between us, thick and suffocating. Three weeks. You're supposed to be careful when you're short, supposed to keep your head down, run out the clock, avoid unnecessary risks. But there is no running out the clock in Afghanistan. Every day you're outside the wire, you're rolling the dice. Marek had been rolling them for eleven and a half months, making it through countless patrols and convoys, and on one of his last rolls, they'd come up snake eyes.

"You should take the rest of the day," I said, meaning it. "Go to the gym, call your family, watch a movie, whatever you need. Nobody's going to judge you for taking time to process this."

Dombroski shook his head firmly. "And do what? Sit in my room and think about it? Stare at the walls imagining him in surgery?" He took another drag. "At least if I'm working, I'm distracted. Besides, someone needs to finish the housing assessment for the Blackhorse project. War doesn't stop because your friend got blown up. The bases keep needing to close. The reports keep needing to be written."

That was the thing about deployment—the relentless forward momentum of it. People got hurt, people died, and you processed it just enough to function, then shoved it down deep into whatever mental compartment you'd created for trauma and kept moving. Because if you stopped to really feel it, to really think about what was happening and what it meant and how fragile everything was, you'd never be able to get back out there. You'd freeze up, shut down, become non-functional. And in a combat zone, non-functional meant dead.

"Well, if you change your mind, let me know," I said, knowing he wouldn't but needing to offer anyway. "Door's always open."

"Thanks, Major. I mean it." He looked at me directly for the first time since I'd walked up, and I saw something raw in his eyes—gratitude mixed with an exhaustion that came from carrying too much for too long. "You're not going to be able to make the basing meeting at IJC today, are you?"

"No, I had to cancel it. Sounds like the Taliban are dug in pretty deep. Doesn't look like the fighting is going to be over any time soon." I paused, already dreading tomorrow. "Hopefully they get their asses kicked by then though. General Washington's replacement is flying in tomorrow afternoon, so I'll be heading up there to pick him up."

"Be careful." The words came out sharp, urgent, and I heard the fear behind them—the plea of someone who'd just lost a friend and couldn't bear to lose anyone else.

"Thanks, I'll be alright." The reassurance felt hollow even as I said it.

"No, I mean it. Really careful." Dombroski stepped closer, his voice dropping lower, more intense. "The route to the Airport is an easy target. The Taliban know our routes and patterns. They've been studying them for years. That's how they got Marek's convoy in Ghazni—they knew exactly where to place the IED, exactly when the convoy would pass. That's how they're getting everyone. They watch, they wait, they learn, and then they strike when you think you're safe."

"I'll be careful," I said again, though the words meant nothing. A promise I had no power to keep.

Because being careful didn't matter. The ISAF advisors had been careful—they'd followed every protocol, varied routes, and maintained situational awareness. Careful didn't save them. Marek had probably been careful too, had probably done everything right. In Afghanistan, being careful was just another way of saying lucky, and luck was a finite resource that eventually ran out for everyone.

We stood there for another moment, two soldiers sharing knowledge we couldn't quite articulate—that we were all living on borrowed time, that every trip outside the wire was a gamble, that the only difference between the living and the dead was often nothing more than being in the wrong place at the wrong time.

"I should get back," Dombroski finally said, crushing his cigarette under his boot. "Those reports won't write themselves."

"Yeah. I've got to grab some documents from HQ anyway." I took one last drag from my own cigarette and ground it out against the concrete wall, the action feeling oddly ceremonial. "Thanks for the smoke."

"Anytime, John. Anytime."

I watched him walk back toward our office, his shoulders carrying the weight of Marek's injury like an invisible burden that grew heavier with each step. Then I turned and headed toward the main headquarters building, my mind already churning through tomorrow's logistics.

The route to NKAIA. The sections of road where traffic slowed, creating natural kill zones for IEDs or ambushes. The tall buildings along the route where snipers could position themselves. Every trip outside the wire required calculating these variables, running the numbers, trying to find the safest route in a place where "safe" was a relative term that meant "probably won't get you killed today."

The sun beat down, baking the ground to powder that hung in the air and coated everything with a fine brown film. Somewhere in the distance, I heard the faint crack of gunfire—could have been training, could have been contact, could have been celebratory fire from Afghans who'd just heard good news. In Kabul, gunfire was ambient noise, as constant and unremarkable as traffic sounds in any other city.

I reached the main headquarters building and pushed through the heavy doors into the air-conditioned interior, the cold air hitting my sweat-soaked uniform like a physical shock. Inside, the building hummed with activity— officers hurrying between meetings, the controlled chaos of a military headquarters managing a war that was supposedly winding down but somehow generated more crises every day.

I grabbed the documents I needed—updated NATO policies, the endless paperwork that kept the drawdown machinery crawling forward— and headed back to my office. The walk gave me time to think, to prepare mentally for tomorrow.

General Washington's replacement. Another trip to NKAIA. Another roll of the dice.

The question I couldn't escape: How many more times could I roll those dice before my luck ran out?

I thought about Jordan back home, about the promise I'd made. I'll come home. But promises in Afghanistan had a way of becoming lies without any warning. You could mean them with everything you had, could intend to keep them with every fiber of your being, and still end up as a casualty statistic because you were in the wrong place at the wrong time.

Survival here required more than skill or training or careful planning. It required luck, and luck was a capricious ally that abandoned people without explanation or apology.

Back in the office, Rod and Bill were deep in conversation about something technical—structural damage of a building at one of the bases we were closing, some mundane detail that nonetheless required precise coordination. The normalcy of it felt surreal after my conversation with Dombroski, after thinking about Marek in surgery.

"Got what you needed?" Bill asked, looking up from his computer screen.

"Yeah. All set." I dropped the documents on my desk, already sorting them mentally into the categories that would determine tomorrow's meetings and next week's briefings.

"You talk to Dombroski?"

"I did. He's holding up. As well as anyone could, I guess."

Bill nodded slowly, his weathered face thoughtful. "Good man. Been through some shit in his career. He'll get through this too."

Would he? I sure hoped so. Everyone had their breaking point, that threshold where accumulating trauma finally exceeded your capacity to compartmentalize and process. Landry had found his in a Dubai airport bar. How many more close calls could Dombroski survive before he found his?

How many more could I?

The rest of the day passed in a blur of meetings and coordination—calls with IJC about rescheduling the basing meeting, emails to Afghan ministry officials apologizing for the delay, reports filed with precise military efficiency despite the knowledge that most of them would be filed away and forgotten.

By 1800 hours, I was exhausted in that bone-deep way that comes from mental rather than physical exertion. I'd spent the day juggling logistics and coordinating schedules, making decisions that affected bases and people across Afghanistan, carrying classified knowledge I couldn't share with anyone, and trying not to think too much about tomorrow's trip.

Back in my quarters, I lay on my cot and stared at the ceiling, listening to the base settle into its nighttime rhythm. Generators humming. Distant conversations in half a dozen languages. The occasional vehicle rolling past. The ordinary sounds of a place at war pretending, for a few hours, to be at rest.

I thought about Marek in surgery, about the surgeons working to save what remained of his leg. I thought about Dombroski against that concrete wall, smoking and staring at nothing, trying to process the impossible. I thought about the two boys executed in Kandahar; their bodies left in the street as a message to anyone watching.

And then, as I always did before a run outside the wire, I thought about tomorrow. The route to NKAIA. The choke points. The blind corners. The hundred ways a ordinary drive could become the last thing you ever did. I'd run the worst-case scenarios so many times they'd become a kind of dark ritual—a way of feeling prepared, of convincing myself that if I'd imagined it, I could survive it.

What I couldn't have known, lying there on my cot staring at a ceiling that offered nothing back, was that somewhere across the city the enemy was already doing the same math. Already choosing the spot. Already setting a trap on the very road I was losing sleep over.

My worst nightmare wasn't a scenario this time.

It was already being built.

Chapter 9: Face to Face with the Enemy

June 11, 2013

The chime of my phone alarm came far too early, dragging me from a restless sleep where I'd spent the night rehearsing every possible disaster that could happen on Airport Road. My subconscious had been particularly creative, serving up scenarios in vivid detail—IEDs, ambushes, snipers, suicide bombers, each one playing out with cinematic clarity before my dream-self died and the scenario reset like a video game loop from hell.

I lay there for a moment in the pre-dawn darkness, my heart already racing, before forcing myself up. Word had spread overnight through official channels that a Special Forces Afghan commando unit had finally broken the Taliban's stranglehold on NKAIA. The enemy forces that had seized control of portions of the base during yesterday's attack had been eliminated—killed in room-to-room fighting that had lasted seventeen hours. Operations were gradually limping back to normal, or at least what passed for normal in Kabul. The airport had reopened. Flights were resuming. The base was functional again, if you ignored the bullet holes and blood stains.

But the fact that the Taliban had managed to penetrate NKAIA's defenses in the first place sent a clear message: nowhere was safe. The enemy could reach out and touch us whenever they wanted, and all our technology and firepower couldn't stop determined men willing to die.

That afternoon, after an uneventful morning, I made my way down the corridor and rapped my knuckles against General Washington's door, trying to push down the anxiety that had been building since yesterday.

"Enter."

I pushed the door open and found him hunched over his desk, reviewing what looked like intelligence reports—probably the after-action analysis from yesterday's attack. His face showed the same fatigue I felt, the accumulation of too many deployments and too many decisions that carried life-and-death consequences.

"Sir, it's ten minutes till. I'm here to take you to NKAIA if you're ready."

He glanced up, and I saw the calculation in his eyes—weighing whether this trip was worth the risk, whether his replacement was important enough to justify driving through a city that had just demonstrated its hostility in spectacular fashion.

"Hi, thanks Maj Spargur. Just give me a second to kit up." He rose from his chair with a grunt and crossed to where his tactical gear leaned against the wall—body armor scarred from months of use, helmet with a spider-web crack in the shell from some previous incident, the accumulated equipment of a deployed officer.

He pulled on his body armor with practiced efficiency, checking straps and adjusting the weight distribution across his shoulders. The Velcro made sharp ripping sounds as he secured the side panels. Helmet next, then tactical gloves and protective glasses. The familiar ritual of preparing for movement outside the wire had become almost meditative—each buckle secured, each magazine checked, each piece of equipment verified. A litany of preparation that was part checklist and part prayer.

Once satisfied with his kit, we headed down the stairs and out toward the motor pool where my Land Cruiser sat waiting in its designated spot, baking in the afternoon sun.

We were maybe twenty feet from the vehicle when the world exploded.

A thunderous explosion tore through the air, so loud and violent it felt like being punched in the chest by an invisible giant. The ground beneath our boots trembled—not shook but actually trembled like a living thing in pain. I felt a light concussive force reverberate through my chest cavity, my lungs compressing, my heart skipping a beat. The blast wave rolled over us with physical weight.

This time General Washington didn't need to identify the sound for me. My short time in Kabul had burned that particular signature into my nervous system.

"VBIED," we said simultaneously, our heads snapping toward the source of the blast like synchronized swimmers. A dark plume of smoke was already beginning to rise in the distance, thick and oily, climbing into the afternoon sky with obscene speed. The mushroom shape it formed was grotesquely beautiful, a perfect pillar of destruction marking where human beings had just been converted into statistics.

"It came from the north," I said, each word feeling heavier than the last. North—the exact direction of our planned route to NKAIA. The universe had just sent us a very clear message about how our day was going to go.

"Yeah, it wasn't very far away either." General Washington's jaw tightened as he stared at the smoke column, his expression shifting through the calculations every deployed officer learned to make: distance, direction, likely target, implications for our mission. He turned back to me, his decision already made before he spoke it aloud. "No use trying to leave until we know what happened. Let's head back to the office and see what Intel reports about this."

We reversed course, leaving the Land Cruiser behind as the distant wail of sirens began to fill the afternoon air. First one, then dozens, building into a symphony of emergency response that would continue for hours. Afghan police. Coalition ambulances. Fire trucks. The familiar soundtrack of catastrophe in Kabul.

The office was already buzzing with activity when I walked in. Everyone had heard the blast—impossible not to, it had rattled windows across the entire base—and they were all clustered around computers, pulling up intelligence feeds, checking secure channels, trying to piece together what had happened from the fragmentary reports coming in.

I took off my gear, knowing with grim certainty it would be hours before we could travel, if we could travel at all today. The body armor came off my shoulders but the weight of anticipation remained, settling into my gut like cold lead. I talked to Rod and Bill, speculating about what could have happened. Was it another ISAF convoy? Civilian neighborhood? A government building? The possibilities were all bad, differentiated only by degrees of horror.

HQ ISAF locked the base down within minutes of the explosion, and for the next couple of hours no one was allowed to enter or leave. The gates were sealed, guards were reinforced, and we all became prisoners of our own security while outside the wire, Kabul bled.

Details of the attack slowly began to unfold through official channels and the informal network of rumors and confirmed reports that any military base develops. Each new piece of information added another layer to the picture, and with each layer, it got worse.

The bombing had taken place at Afghanistan's Supreme Court, situated on the route between HQ ISAF and NKAIA—exactly on our planned path. It was the deadliest attack in Kabul since December 6, 2011, when a suicide bomber had killed over 60 people during the Ashura commemoration.[3] The Taliban had learned from that success and refined their tactics.

This time, a Taliban suicide bomber had driven his SUV—packed with an estimated 1,500 pounds of homemade explosives—and specifically

---

[3] Roggio, Bill. *"Suicide Bomber Kills Scores in Attack at Kabul Mosque."* FDD's Long War Journal, 6 Dec. 2011, longwarjournal.org/archives/2011/12/suicide_bomber_kills_61.php.

targeted the buses filled with court workers who were about to leave for the day. The timing had been perfect, calculated to maximize casualties. The explosion occurred at 1600 hours, exactly when the Supreme Court employees finished their shifts and gathered at the designated pickup point.

Reports came through that at least four buses and more than ten civilian vehicles had been destroyed in the attack. The twisted metal and charred remains of the buses painted a picture I tried not to visualize but couldn't help imagining—bodies torn apart, burned beyond recognition, scattered across pavement that had been superheated by the blast.

The casualty numbers climbed as the afternoon wore on: 17 dead and over 40 seriously injured, all civilians.[4] Judges, clerks, translators, janitors—people just trying to get home to their families. The Taliban's message was brutally clear: the Afghan court system and its judges must pay for obeying the Western world, for applying laws that didn't come from their interpretation of Islamic jurisprudence, for cooperating with an occupying force.

Just before 1800 hours, General Washington appeared in our doorway. His face carried that particular expression of grim determination I'd learned to recognize—the look of a commander who'd weighed all the factors and decided the mission had to proceed despite the risks.

"Let's go, Spargur. Roads have been cleared for travel."

Dinner tonight would have to wait, I thought, my stomach already tight with anxiety. "Yes sir, I'll be right there."

I kitted up for the second time that day, strapping on body armor that felt heavier than it had this morning, securing the helmet that suddenly seemed inadequate for the threats we'd face, checking my M4 with hands that were steadier than I felt. This time there were no interruptions as we headed back to the motor pool, no explosions to delay us, just the fear of not knowing what lay ahead.

---

[4] Nordland, Rod. "Taliban Bomb Attack Kills Court Workers in Kabul." The New York Times, 11 June 2013, www.nytimes.com/2013/06/12/world/asia/afghanistan.html.

After radioing our departure off base—calling sign "Juliett Sierra" to notify headquarters we were leaving the wire—we headed onto the streets of Kabul. Evening traffic was eerily calm, unlike the usual congested bumper-to-bumper chaos. The blast had cleared the roads more effectively than any traffic control, fear keeping civilians at home behind locked doors.

Taking the north exit of the Massoud roundabout onto Airport Road, I noticed debris scattered everywhere—broken glass glittering in the fading sunlight like diamonds, chunks of concrete and metal scattered across the road and sidewalks like offerings to some violent god. The street had been swept enough to allow passage but not cleaned, the evidence of destruction deliberately left visible.

Directly to the east, behind a row of trees that had been stripped of leaves by the blast, sat the Supreme Court building. Scorch marks blackened the area near where the SUV had detonated. The building itself showed damage—blown-out windows, pockmarked walls, a section of façade that had partially collapsed. The acrid smell of explosives still hung in the air, mixed with something else I didn't want to identify but recognized anyway: burned flesh, that sweet-sick odor that embedded itself in your memory forever.

By the time we reached the south gate of NKAIA, the sun had sunk low on the horizon, painting the sky in bands of deep orange and purple. Dusk was settling over Kabul like a heavy blanket, softening the edges of destruction with beauty that felt obscene given what had happened hours earlier. The haunting call to prayer—one of the final Muslim prayers of the day—echoed across the city from unseen minarets, the voices layered and ethereal in the cooling air. Allahu Akbar. God is great. The words carried across dusty streets and through blast-damaged buildings, a reminder that faith persisted even when everything else fell apart.

As we rolled up to the south NKAI base gate entrance, I noticed immediately that the gate was sealed shut, a massive metal barrier designed to stop vehicle-borne threats. Standing just outside the perimeter wall, an Afghan National Army sentry had his back turned completely toward us, his rifle slung over his shoulder as he faced the concrete barrier in prayer.

His forehead touched the ground in prostration, oblivious to our approach, completely absorbed in his spiritual obligations.

Geez, I thought, gripping the steering wheel a little tighter. The best time to attack this place would be during prayer time. The tactical vulnerability was glaring—this guard didn't even seem to register our presence, let alone identify whether we posed a threat. We could have been anyone rolling up to that gate: legitimate personnel, or a suicide bomber timing his approach to coincide with the guard's prayers. He wouldn't have known until it was too late.

"Go tell this guy to open the gate," Washington said, his voice barely audible over the amplified prayers still reverberating through the air from nearby loudspeakers.

"Yes, sir." I grabbed my M4 from beside my seat and pushed open the door, boots hitting the coarse ground with a crunch that went unheard beneath the prayer call.

"Excuse me," I called out firmly as I approached the guard station, trying to cut through the wall of sound still pouring from the loudspeakers. The ANA soldier remained motionless, deep in prayer, his entire body oriented toward Mecca. I moved closer, now standing just a few feet away at the edge of his small concrete shelter. "As-salamu alaykum," I tried again, louder this time, offering the traditional Islamic greeting.

Still nothing. He continued his prayers, forehead pressed to the ground, hands positioned precisely, completely absorbed in his ritual communication with God.

I shifted my weight, glancing back at the Land Cruiser where I could see Washington silhouetted in the gathering darkness, probably getting increasingly frustrated about keeping his replacement waiting for hours. I guess I'll wait, I thought, though I knew the general was already irritated about the delays. The seconds crawled by—fifteen, thirty, forty-five—until finally the prayer call faded into silence, the last echoes dying across the city.

The guard straightened slowly and turned around, his face registering surprise when he saw me standing there like an apparition that had materialized during his prayers. His expression shifted quickly from surprise to something less welcoming—clear displeasure that I'd been standing there, waiting to interrupt the moment his prayers concluded. His jaw tightened, and his eyes held that particular look of resentment that came from having your religious observance treated as an inconvenience by infidel foreigners.

"Salam," I said quickly, not wanting to waste any more time or escalate whatever irritation he felt. "I'm from HQ ISAF and I need you to open the gate. I have business inside."

He studied me for a long moment, his eyes moving from my face to my weapon to the vehicle behind me, conducting his own threat assessment. Then he responded in heavily accented, fractured English that came out slowly and deliberately, as if each word had to be translated in his head before being spoken.

Due to all the recent terrorist activity, he explained with careful deliberation, this gate would remain closed until daylight the next morning. Escalated security protocols had been implemented after yesterday's attack on the base. However, he added with a slight tilt of his head toward the north, the guards at the north gate might be able to let us through. Maybe. No promises.

"Tashakur," I said, offering my thanks in Dari before jogging back to the vehicle, already dreading Washington's reaction.

I slid back into the driver's seat and relayed the information to Washington, who let out a slow breath through his nose—the kind of controlled exhalation that suggested he was working very hard not to express his frustration more colorfully.

"I've never been to the north gate before," I admitted, already dreading what came next. "And I have no idea how to get there."

"We'll figure it out." Washington's tone left no room for debate, squashing any discussion about whether this was a good idea or whether we should just call it a day and come back tomorrow in daylight when the south gate would be open. "Just follow the east perimeter of the base, then hook around to the north. Drive."

I put the Land Cruiser in gear and pulled away from the closed gate, heading into the deepening twilight with only a vague sense of direction and the general's confidence to guide us. His plan sounded easy enough in theory, but there were two critical problems rapidly becoming apparent: we were losing daylight fast, and there were no easy roads that simply hugged the base perimeter like he'd suggested.

I pointed the Cruiser eastward and drove into uncharted territory, through neighborhoods I'd never seen. Within a few minutes of zigzagging back and forth on what seemed like a maze of narrow streets, I was completely lost. With no map, no GPS, and fading landmarks as the sun disappeared below the horizon, I was struggling to maintain any sense of direction. The narrow streets all looked identical in the gathering darkness—high mud-brick walls, metal doors set into them, shadows deepening in every corner, the architecture of a city that had learned to hide its secrets behind barriers.

"Turn here," Washington said, pointing to a narrow road heading north that looked like it might parallel the base perimeter.

I turned cautiously, my headlights cutting through the gloom. Running parallel to the road on the right was a typical ten-foot-high mud brick wall—the kind that protected Afghan family compounds, their height designed for privacy and security. To my left was a drainage ditch, also typical of the local landscape, running alongside the road. Further from the city center though, this ditch didn't have a metal grate covering it like the ones downtown had—just an open trench that could easily swallow a tire or trap a vehicle.

As I peered down the road ahead, squinting against the darkness, I spotted a figure on the left side of the street. An Afghan man with an AK-47. Behind him stretched an old metal arm gate across the road, its tip reaching almost to the end of the mud brick wall. Behind the gate, the wall ended abruptly and opened up into what appeared to be a large open field—darkness stretching away into nothing.

This is a strange place for a checkpoint, I thought as I eased the Land Cruiser closer to the unusual control point, my instincts already screaming warnings my conscious mind hadn't fully processed yet. The structure looked hastily assembled, improvised, nothing like the fortified gates and professional checkpoints we'd passed earlier. There were no barriers, no sandbags, no proper guard tower—just a man and a gate in the middle of nowhere.

I brought us to a stop about twenty-five yards out, leaving enough distance to maneuver if we needed to reverse quickly. The tactical part of my brain was already calculating escape routes, noting the ditch to our left that would prevent easy maneuvering, the wall to our right that blocked any possibility of going that direction, the narrow road behind us that would require backing up.

A feeling of uneasiness crept through me like cold water trickling down my spine, starting in my chest and spreading outward through my limbs. Something was wrong. Every instinct I'd developed over my time in Afghanistan was screaming at me, sending alarm signals that bypassed conscious thought and went straight to the primal part of my brain responsible for survival.

In the rapidly dimming light, I could make out details of the single Afghan man standing near the gate. He appeared to be in his thirties, with a thick black beard and a black turban wrapped tightly around his head— not the typical green or brown or white turbans I'd seen on most Afghans, but black. Taliban black. He wore a beige tunic and loose trousers, paired with a black vest that hung open at the front. Everything about his appearance seemed off—not quite military, not quite civilian, something else entirely.

The man's eyes widened slightly when he first noticed us, a flicker of surprise crossing his face before his expression settled into something more neutral. I watched him carefully, expecting the routine that usually followed at legitimate checkpoints: a cursory inspection, maybe a quick verbal greeting, a check of our vehicle identification, then the wave through. Instead, he reached into his vest and pulled out a cell phone.

My stomach dropped.

He began speaking rapidly to someone on the other end, his eyes darting toward us every few seconds but never making direct contact. Not once did he acknowledge our presence with a nod, a wave, or any gesture of recognition. His body language was all wrong—not the respectful deference Afghan National Police or Army usually showed coalition forces, nor the professional efficiency of legitimate security personnel. He looked at us the way you'd look at an obstacle that needed to be managed, or prey that needed to be trapped.

A pit began forming in my stomach, hard and cold, a visceral dread that comes from recognizing danger before your conscious mind can articulate why. The demeanor of this man was fundamentally different from every Afghan checkpoint guard I'd encountered during my deployment. Private security contractors and ANA soldiers were typically eager—sometimes overeager—to assist coalition forces. They'd salute, smile, offer chai, anything to demonstrate cooperation and establish rapport. This man looked at us the way you'd look at a problem to be solved, or targets to be eliminated.

"Should I put it in reverse and get us the hell out of here?" I asked Washington, unable to keep the concern from creeping into my voice. Every instinct I had was screaming at me to back up, turn around, find another route. My hands were already tense on the wheel, ready to jam the transmission into reverse. Please say yes. Please.

"No, we're going through this gate," he replied flatly, his tone suggesting my question had been not just unnecessary but borderline irritating, as if I was being overly cautious about nothing.

But it wasn't nothing. My gut knew it wasn't nothing.

By this time the Afghan had finished his phone conversation, lowering the device slowly. But rather than approaching us or opening the gate, he simply stood there in the deepening twilight, his body angled toward the outer edge of the wall, his gaze fixed on the dark field beyond the perimeter. Waiting. For something. Or someone.

The silence stretched, broken only by the idling engine of our Land Cruiser and the distant sounds of the city—dogs barking, a siren wailing, the ambient noise of life continuing somewhere far away from this isolated road where we sat like sitting ducks.

General Washington shifted in his seat, his patience clearly running out. He leaned forward and began waving his hands emphatically through the windshield, as if he could physically drag the man's attention back to us through sheer force of will. "Hello! Open the gate!" he shouted toward the bulletproof glass, his voice muffled but audible, frustration bleeding into his command tone. He jabbed his index finger up and down in an exaggerated pointing motion.

The man remained motionless, a statue in the fading light.

Now visibly upset, Washington started pounding his fist against the windshield—thump, thump, thump—the dull impacts reverberating through the cabin like a heartbeat of frustration. He pointed aggressively at the small, laminated placard affixed to the passenger corner of the windshield. The sign, printed in Arabic and English with official NATO emblems, clearly identified this as an ISAF vehicle with authorization to travel anywhere in Afghanistan. It was supposed to be our golden ticket, our diplomatic passport, our pass through any checkpoint in the country.

Still nothing. No movement. No acknowledgment that two coalition officers sat twenty-five yards away in an official vehicle, waiting to be let through. The man might as well have been carved from stone.

Fear began to take hold, tightening around my throat like invisible hands. I glanced at Washington, searching his face for some sign that he shared my growing alarm, that his combat experience was triggering the same warning bells mine was. He's got multiple combat tours, I reminded myself, trying desperately to ease the anxiety clawing at my thoughts. Surely he knows what he's doing. Surely he sees something I don't. Trust the senior officer. Trust his experience.

But the pit in my stomach told me otherwise. Sometimes rank and experience didn't matter. Sometimes your gut knew things your brain hadn't caught up to yet.

Then, without warning, movement shattered the stillness.

Two more Afghan men materialized from the shadows around the corner of the mud brick wall, their forms emerging like wraiths from the darkness itself. Both carried AK-47s, the distinctive curved magazines unmistakable even in the failing light, the weapons held casually but ready. For the first time since we'd arrived, the man who'd been standing motionless at his post stirred. He stepped forward deliberately, greeting the newcomers with familiar ease—the body language of people who knew each other well, who'd worked together before.

The three men huddled together just feet from our front bumper, their heads bent close in urgent conversation. Their voices were low, inaudible over the idling engine, but their body language spoke volumes—quick gestures toward us, pointed glances in our direction, nods of agreement about something I desperately wanted to know but terrifyingly didn't. After perhaps thirty seconds that felt like thirty minutes, they broke apart and began walking toward our vehicle in a loose formation, all three wearing black turbans, all three with AKs held casually but ready, their movements coordinated like a practiced team.

My blood turned to ice.

Holy shit. We've stumbled into some kind of Taliban nest.

My fear crystallized into full-blown panic as they began to circle the Land Cruiser with predatory slowness, like wolves sizing up cornered prey. They peered through the windows with deliberate thoroughness; their faces mere inches from the bulletproof glass, studying every detail of the interior with the clinical assessment of professionals planning an operation. One of them—a man with piercing black eyes set in a leathery, scarred face beneath his beard—stopped directly at my window.

He stared at me with an intensity that felt physical and visceral, an aggressive intrusion into my personal space. I could feel the hatred radiating from him through the heavily armored door like heat from a furnace, could sense the violence coiled just beneath his weathered exterior, barely contained. His eyes held stories I didn't want to read—how many Americans he'd killed, how many attacks he'd planned, how many IEDs he'd buried that had torn coalition soldiers apart.

His gaze shifted methodically, cataloging everything inside the vehicle with practiced precision. The M4 assault rifles resting against our seats. The 9mm handguns strapped to our hips. Our body armor. Our positioning.

The radio. Our reaction. He was sizing us up, running calculations behind those cold eyes. Threat assessment. Target analysis. Vulnerability identification. This was a man who'd done this before, who knew how to evaluate targets and exploit weaknesses.

"Sir, this doesn't look good," I said, hearing the desperation creep into my own voice despite my best efforts to maintain military composure. My training said stay calm, project confidence, show no fear. But my survival instinct said get the fuck out of here right now. "Let's try a different route. We should leave. Now. Right now."

"No. We're going through this gate," Washington said with unwavering firmness, as if the danger unfolding around us was merely a bureaucratic inconvenience to be overcome through sheer force of will. His jaw set with that stubborn determination I'd seen before—the refusal to be deterred, to back down, to admit that maybe this was a bad idea. He leaned forward again, his face flushed with frustration and probably adrenaline though he was channeling it into anger rather than fear. "Open the damn gate!" he shouted, slamming his fist against the windshield and jabbing his finger at the ISAF identification placard.

This time, all three men moved in unison to the passenger side, drawn by the noise and movement like sharks to blood in the water. They clustered near Washington's door, examining the sign with exaggerated interest, their heads tilted as they spoke among themselves in rapid Pashto. Something about their deliberate movements, the theatrical quality of their inspection, set off every alarm bell in my head. They weren't trying to read our credentials—they were positioning themselves.

"What the hell is wrong with these guys," Washington muttered, his patience finally exhausted, his voice carrying the edge of someone about to do something irreversibly stupid. "I'll open the damn gate myself."

And then, in the moment that would define everything that followed, in the split-second that separated one version of reality from another, I watched—helpless, disbelieving, my mind unable to process what I was seeing—as his hand moved to the door panel.

No.

The soft click of the unlock button echoed like a gunshot in my ears, reverberating through my skull with the finality of a judge's gavel.

No, no, no.

Then he reached for the heavy-duty four-inch door deadbolt, the auxiliary lock that provided an extra layer of security beyond the standard mechanism. His fingers grasped the metal lever. I wanted to scream, to reach across and stop him, to grab his hand, but I was frozen—paralyzed by disbelief that this was actually happening.

The deadbolt was the only real barrier between us and them. The only thing keeping us safe in this up-armored Toyota that could withstand bullets but not determined men if they got inside. And now he was voluntarily removing it, opening himself up to the evil that circled us like sharks scenting blood in dark water.

My mind flashed unbidden to my son's face. Jordan. His smile. The promise I'd made before deploying: I'll come home. I promise, buddy. I saw him suddenly as if projected on the windshield—older, haunted, watching grainy footage on the news of his father being dragged from a vehicle, captured and beheaded in some godforsaken Taliban cave, his final moments broadcast to the world as propaganda. I saw my mother opening the door to find uniformed officers on her porch. I saw the funeral. The flag. The emptiness.

I'm about to break my promise.

The door cracked open—just a sliver of space between the frame and the cab, but in that instant, everything changed.

The three men exploded into action with terrifying coordination.

One lunged forward and grabbed the door's edge, wrenching it wider with surprising strength, muscles corded in his forearms as he hauled against Washington's resistance. The other two descended on him simultaneously, their hands clawing at his right arm which still clutched desperately at the inner door handle. It became a grotesque tug-of-war: one man pulling the door open while the other two physically hauled the General toward the opening, trying to drag his entire body out of the vehicle like predators pulling prey from a den.

Washington's boots scraped against the floor mat as he fought to hold his ground, his face contorted with effort and growing terror—the realization too late that he'd made a catastrophic error in judgment. The professional military bearing that usually defined him shattered completely, revealing the raw animal fear underneath.

My hand shot down to the 9mm pistol holstered on my hip, fingers wrapping around the grip with practiced efficiency. We're going to have to shoot our way out of here. The thought came with crystal clarity even as adrenaline flooded my system, my heart hammering against my ribs so hard I could hear it in my ears. I drew the weapon, my movements smooth despite my hands wanting to shake, muscle memory from combat training taking over.

"Get us the fuck out of here, John!" Washington screamed, and I'd never heard that tone in his voice before—raw panic stripped of all command authority, just pure animal fear. His legs kicked wildly as he struggled against his attackers, trying to create any separation, any breathing room. One of his boots connected with something solid and I heard a grunt of pain from outside.

I yanked the 9mm free from its holster and quickly chambered a round, the mechanical slide-rack both familiar and terrifying in this context—that distinctive *chak-chak* sound that meant the weapon was hot, ready to end lives. I searched frantically for a clear shot, swinging the muzzle toward the passenger side where the struggle continued, but there was nothing. Washington had been dragged so far out of the vehicle that his large frame now completely filled the doorway, a writhing mass of limbs and bodies locked in desperate combat. Beyond him, the attackers' faces appeared and disappeared in the struggle—bearded, intent, determined. I couldn't fire without hitting him.

The tactical calculus was simple and horrifying: if I shot now, I'd kill my commanding officer. If I didn't shoot, they'd drag him out and I'd be alone, surrounded, outnumbered, and they'd come for me next.

Then something shifted in the struggle. Now, having been pulled entirely out of the SUV, Washington somehow began to gain the upper hand through sheer desperate violence. His left hand shot out and clamped onto the passenger doorframe with iron grip, fingers digging into the metal trim with the strength of a drowning man grabbing a rope. With his feet planted solidly on the ground, he twisted his body with violent force and unleashed a series of devastating kicks and punches—pure survival instinct and rage combined into explosive action.

His boot connected solidly with one attacker's groin, the impact producing a painful moan. The man's head snapped forward, his body crumpling. Washington's fist crashed into another's temple with a sickening thud, a blow delivered with every ounce of strength and all the force of desperation. The impact knocked the second attacker backward, off-balance. The blows were brutal, primal, ugly—not the controlled techniques taught in combat training but the wild haymakers of a man fighting for his life, and effective enough to knock two of them backward, giving him the precious seconds he needed.

With both hands now free, he grabbed the door and hauled himself back inside with a roar of effort, his face a mask of fury and terror combined. But the third man—the one who'd been gripping the door—renewed his efforts to keep it open, his fingers wrapped around the edge, pulling with everything he had.

"Go, go, go!" Washington roared, back in his seat but still fighting to pull the door closed as the attacker maintained his grip, muscles straining, teeth bared in effort.

I knew exactly what I had to do. There was no time for hesitation, no room for second-guessing, no opportunity to weigh options or consider consequences. Action was survival.

My right hand shot to the gear selector, slamming the transmission into Drive with enough force that the mechanism clunked in protest. My left foot came off the brake as my right stomped the accelerator to the floor in one fluid motion, every pound of pressure my leg could generate. The Land Cruiser's powerful engine roared to life with a guttural bellow that drowned out the sounds of struggle. The tires spun momentarily on the loose dirt, throwing up a spray of dirt and gravel before the treads found traction and the vehicle lurched forward violently.

The sudden surge of momentum did exactly what I'd hoped—it shattered the Taliban fighter's grip on the door. His fingers slid off the metal edge as physics overtook determination, his feet stumbling as he desperately tried to keep pace with our rapidly accelerating vehicle. I saw him in the side mirror, his body toppling over as he lost the race, arms windmilling as he fell.

Click.

The sound of the door latching—that simple mechanical engagement of metal on metal—was the most beautiful sound I'd ever heard. It meant safety. It meant survival. It meant the barrier between us and them was restored.

General Washington collapsed back into his seat, his chest heaving with ragged breaths, his hands still gripping the door handle with white-knuckled intensity as if he didn't quite believe it was closed. His right tactical glove was gone, pulled off in the struggle.

Darkness had descended now, swallowing the landscape in shadow. I quickly glanced over my shoulder through the rear passenger windows and felt my heart drop into my stomach. Three shadows were scrambling back along the wall toward us, regrouping with frightening speed. They weren't giving up. This wasn't over.

Without thinking, operating on pure instinct and survival training, I slammed the Cruiser into Reverse and hammered the accelerator again. Please God, I prayed silently as the vehicle shot backward, engine screaming in protest, don't let me fall into that ditch. The drainage ditch was somewhere to my left, invisible in the darkness, ready to swallow a tire and trap us here like animals in a snare.

Cranking the steering wheel hard counterclockwise, I yanked the Cruiser into a violent pivot, the tires shrieking as rubber fought dirt for traction. The whole frame lurched, suspension groaning under the sudden shift as the front end whipped around in a tight, brutal arc. For an instant the world narrowed to the side mirror and the cluster of figures behind—our attackers scrambling towards the wall, aware of the two-ton battering ram bearing down on them. Then the gap closed.

The front passenger quarter panel struck first, followed a heartbeat later by the reinforced bumper slamming into the fighters with a bone-jarring finality. The impact sent them tumbling like rag dolls, bodies flung aside by sheer momentum. A sickening thud reverberated through the steering wheel and up my arms—metal colliding with flesh and bone at speed.

I spun the steering wheel again, this time centering the vehicle in the middle of the narrow street, and kept my foot buried in the accelerator. The transmission screamed in protest—a high-pitched mechanical wail that suggested I was about to destroy the vehicle's drivetrain—as we continued our escape driving backward down the road at dangerous speed. The engine wasn't designed for this kind of sustained reverse operation at high RPM, but I didn't care. Distance. We needed distance. Space between us and the men who'd tried to drag us from our vehicle.

Using the wall to my right as a visual guide, I focused on keeping us within three feet of its edge while maintaining straight wheels, a feat made nearly impossible by the speed and the darkness and the adrenaline coursing through my system making my hands tremble on the wheel. In the pitch black, with only my mirrors and the faint glow of the reverse lights to navigate by, it was like threading a needle while riding a roller coaster. But I knew if I could stay parallel to that wall, if I could keep the wheels straight and avoid overcorrecting in panic, we had a chance of making it out.

I stole a quick glance forward through the windshield, my headlights now illuminating the scene we were fleeing. One Taliban fighter I'd struck was moving, slowly pushing himself up from the ground where he'd fallen, dazed but alive. The other two had recovered and were actively trying to retrieve their AK-47s which had been scattered during the chaos, their hands reaching for weapons in fluid practiced motions. Any second now they'd have those rifles in hand and we'd be taking fire.

Then, suddenly, the wall on my right side ended—the edge of the compound materializing out of darkness without warning. I slammed on the brakes, the Land Cruiser sliding sideways on the loose gravel with a sickening grinding sound that suggested I'd just taken years off the brake pads and tires. Before we'd fully stopped, while the vehicle was still skidding, I cranked the wheel hard over, shifted back into Drive, and punched the accelerator once more. The vehicle launched forward, tires biting into the earth as we rocketed back toward the distant glow of city lights on the horizon.

Still lost, still disoriented, navigation now the least of my concerns, but pumping with pure adrenaline that made everything feel both hyper-focused and surreal simultaneously. I drove as fast as the narrow road and darkness would allow, my eyes flicking constantly to the rearview mirror, expecting at any second to see headlights—a Toyota Hilux packed with armed fighters, racing up behind us with AK-47s bristling from the windows like spines on a porcupine.

"HQ, this is Lima Whiskey, how do you copy over?" General Washington's voice cut through the cabin, strained and urgent, the professional military bearing struggling to reassert itself over the tremor of fear still audible underneath. He held the radio handset to his mouth with his bare right hand, his thumb pressing the transmit button.

Static. Nothing but electronic hiss and pop crackled across the speaker, the white noise of failed communications.

"HQ, do you copy over?" He tried again, louder this time, as if volume might somehow force the signal through whatever was blocking it. The desperation bled through despite his best efforts to sound calm.

More static.

"Pull over here," Washington said, his voice trembling in a way I'd never heard before—not angry, not commanding, just shaken to its core like a building after an earthquake, the foundation cracked but not yet collapsed. "We've got to report this and I can't get this piece-of-shit radio to work."

I eased off the accelerator and guided the Cruiser to the side of the road, choosing a spot where I had decent visibility in both directions—could see headlights approaching from either way, could maneuver if we needed to run again. Even as I shifted into Park, my eyes remained locked on the rearview mirror, searching the darkness behind us for any sign of pursuit.

My hands were trembling on the steering wheel, a fine tremor I couldn't control no matter how hard I gripped. My leg muscles twitched involuntarily, still flooded with adrenaline. Every fiber of my body remained on high alert, my nervous system refusing to accept that we'd escaped, that the immediate danger had passed.

The engine idled roughly in the silence that followed, and for a moment, neither of us spoke. We just sat there in the darkness, two officers in a disabled vehicle in hostile territory, trying to process what had just happened and figure out our next move.

"Where's the damn console light switch?" Washington said finally, breaking the silence, his voice more controlled now but still carrying that edge of barely suppressed panic. He was searching for the controls, hands moving across the dashboard.

"Here you go," I said, flicking the switch and giving him enough light to see the VHF radio controls. Nothing he tried seemed to work. Static was the only response to every frequency adjustment, every attempt to reach headquarters.

"Let me try my cell phone," he said at last, pulling a silver Nokia from his pocket—the old brick-style phone that had been standard issue before smartphones took over. He dialed once, twice, three times, each time failing to get a signal, his frustration mounting with each failed attempt.

"You're not going to be able to get through," I said, knowing what was likely causing the interference, my voice coming out flatter than I intended. As a former Space Defense Director with the Air Force, I was very familiar with Electronic Warfare and the effects it caused. After my meetings with the CIA, I had a very good idea about where those EW systems operated from. It made perfect sense that on a night like tonight—with so much Taliban activity during the week, attacks ramping up, the spring offensive in full swing—our covert forces would be conducting electronic warfare operations near the airport, jamming communications to disrupt enemy coordination.

"Yeah, why's that?" Washington asked, looking at his phone screen as if it had personally betrayed him.

"Our guys are jamming the signal," I explained, trying to keep my voice professional despite wanting to scream because they're running operations while we're driving around lost in Taliban-controlled neighborhoods. "Likely performing some sort of EW ops near the airport to disrupt Taliban communications. Cell phones, radios, everything's getting jammed in this area."

He paused, processing that information, probably running through the same calculations I was: we were on our own, cut off from support, lost in hostile territory with no way to call for help. Then his mind shifted back to the mission with the single-minded focus of a commander who'd decided retreat wasn't an option.

"Alright, let's make our way back to HQ," he said, his voice carrying that tone of finality that suggested the discussion was over. "I'll report this when we get back. I want you to get Rod and come back to pick up General Williams."

The words hit me like a physical blow.

What. The. Fuck.

We'd narrowly escaped getting kidnapped or killed because of his horrible decision to open that door, to expose us to an enemy that had been circling us like predators, and now he wanted to send me back onto these roads late at night in a city where the Taliban had been wreaking havoc all week? After what we'd just experienced? After coming within seconds of being dragged from our vehicle?

Rage flooded through me, hot and immediate, burning away the fear and replacing it with white-hot anger. It was reckless. It was stupid. It was completely unnecessary. We'd already risked our lives once tonight and accomplished nothing except almost getting captured. And now he wanted me to do it again?

I took a deep breath and slowly exhaled before speaking, trying to force my voice to remain calm and professional when every cell in my body wanted to scream at the absurdity of what he was proposing. "Sir, can't General Williams just stay overnight at NKAIA? They've got facilities there he can sleep in—transient quarters, a DFAC, everything he needs. I'm sure he's probably heard about the attack today at the Supreme Court and would understand the delay."

"No," Washington said flatly, his command voice reasserting itself now that we were out of immediate danger. "I told him we'd be there tonight to get him. Rod should know how to get to the North Gate. Let's get going."

I was furious. Beyond furious. I took another deep breath, held it for a three-count, and slowly exhaled, trying to process the order I'd just been given. Every instinct screamed at me to refuse, to say that I wasn't risking my life again for a mission that could wait until morning when we had daylight and better security.

But you don't refuse orders. Not in the military. Not when a senior officer tells you to do something, even when that something is breathtakingly stupid and unnecessarily dangerous. The machine doesn't work if people start making their own decisions about which orders to follow.

I eased back onto the road without another word, focusing all my attention on getting us safely back to HQ ISAF. Keeping my eyes on the twinkling lights of downtown Kabul in the distance, I headed south and then made my way west until I bisected Airport Road. From there it was familiar territory—a route I'd driven dozens of times, landmarks I recognized even in darkness. An easy drive back to the gates of HQ ISAF if you ignored the bombing site we'd pass, the bodies that had been there hours ago, the knowledge that every intersection could hide an IED.

After what seemed like the longest trip of my life—thirty minutes that felt like thirty hours—I parked in the motor pool and cut the engine. The sudden silence was deafening after hours of engine noise and rushing blood in my ears.

"Tell General Williams I'll meet him at 0700 tomorrow morning at the office," Washington said, opening his door with his bare right hand, his tactical glove gone forever, pulled off by Taliban fighters who'd tried to drag him to his death. "Make sure you guys help him take his gear to the barracks. He won't know where to go. Goodnight."

Goodnight. The word hung in the air, absurd and offensive.

I didn't say a word. Couldn't trust myself to speak without saying something that would end my career. This had been anything but a good night. This was the worst night of my life. Thank God it didn't end up being the last night of my life.

I crawled out of the Cruiser, my legs shaky, my body feeling like it weighed a thousand pounds. All the adrenaline from earlier had been depleted, burned through, leaving behind nothing but exhaustion so profound it felt like gravity had doubled. I slowly made my way toward the barracks, stopping off at the bathroom along the way.

The fluorescent lights were harsh after the darkness outside. I leaned over the sink and splashed cold water over my face, the chill shocking against my overheated skin. Looking up into the mirror, I saw water mixed with tears running down my face. When had I started crying? I didn't remember. My eyes were red, my face was pale, and I looked ten years older than I had this morning.

"Pull yourself together, John," I told my reflection, grabbing a paper towel and wiping my face dry with rough, angry motions. But the face staring back at me didn't look convinced. That face had seen too much. Had come too close to the edge.

I found Rod on his cot, headphones on, watching a movie on his laptop—some action film where the good guys always won and nobody actually died. He looked up from his screen when I entered, surprised to see me there at such a late hour, his face shifting from relaxed to concerned in the span of a heartbeat when he saw my expression.

"Sorry to bug you Rod," I said as he pulled off his earphones, sitting up on the edge of his cot. "We've got to go to NKAIA and get the new General."

"I thought you were taking General Washington to get him," he said, looking confused and maybe a little worried. "Wasn't that the plan?"

"Yeah, I did and it didn't turn out so well. I'll tell you the story along the way." My voice came out flat, emotionless, the tone of someone who'd shut down their feelings to keep functioning. "Get dressed and kitted up. I'll meet you at the motor pool in 10 minutes."

Ten minutes later, Rod climbed into the passenger seat of the Cruiser, his C7 assault rifle already locked and loaded, a round chambered and ready. He looked at me expectantly as I started the engine, probably wondering why I looked like I'd aged a decade since breakfast.

"So, what happened?" he asked as we rolled out of the motor pool and headed toward the gate.

I took a breath and began telling the story, keeping my voice level and factual, narrating the events like I was filing an after-action report rather than describing the most terrifying experience of my life. "We drove to the South Gate of NKAIA but it was closed for the night because of yesterday and today's attacks. The guard said we might be able to get in through the North Gate, so Washington told me to drive around the perimeter to find it."

"Okay…" Rod said slowly, already sensing this was going somewhere bad.

"Problem is, I've never been to the North Gate. Don't even know where it is. And by this time, it's getting dark—really dark. No streetlights, unfamiliar roads, everything looking the same." I paused, checking my mirrors out of habit even though we were still on base. "We ended up on this narrow road, real isolated area, no other vehicles. Then I see this guy standing in the middle of the street with an AK-47."

"ANA?" Rod asked hopefully, though his tone suggested he already knew the answer.

"That's what I thought at first. But when we drove closer, I could see he wasn't wearing a uniform. He was wearing a black turban—Taliban black— had this long beard, and he just… he didn't act right. He saw us and instead of opening the gate or checking our credentials or any of the normal checkpoint procedures, he pulled out a cell phone and started making calls while staring at us."

"Shit," Rod muttered, his hand unconsciously moving to check his rifle.

"Yeah. Then two more guys with AKs showed up out of nowhere. All three of them had black turbans. Taliban turbans, Rod. They started circling the vehicle, looking at our weapons, staring through the windows, just…

sizing us up like we were prey and they were deciding how to take us down." I gripped the steering wheel tighter, the memory making my pulse quicken even now. "I told Washington we should reverse out of there. I practically begged him. He said no. He wanted to go through that gate."

"What?" Rod turned to look at me directly, his eyes wide. "Are you serious? With a potential Taliban team surrounding you?"

"Dead serious. To me, their calculus was obvious: alive, we had value. Dead, we were just bodies. We were high value assets that they desperately wanted. How he failed to recognize that, I have no idea. Then he did something that I still can't believe, something so incomprehensible I'm still processing it." I shook my head. "He unlocked his door and opened it."

"He what?" Rod's voice rose with disbelief.

"Opened his door. Just opened it right into their hands. Walked right into their trap. They grabbed him immediately, tried to take his gun and drag him out of the vehicle. It turned into a full-on brawl right there in the doorway—three Taliban fighters trying to pull him out while he fought to stay in."

Rod sat in stunned silence for a moment, trying to process what I'd just told him. "Jesus Christ, John. What did you do?"

"I drew my 9mm, chambered a round, tried to find a shot, but I couldn't get a clear line— Washington was right in the middle of it all, and if I fired, I would have hit him. Somehow, he fought them off long enough to get back in the vehicle. The second he did, I punched it and got us the hell out of there.

"And now he's sending us back out there?" Rod's voice rose with disbelief and anger. "After what just happened? Is he crazy?"

"That's what I said. Well, not exactly those words because he's a General, but yeah—that's essentially what I was thinking." The bitterness bled through despite my best efforts to keep it professional. "He wants General Williams picked up tonight, no matter what. Says you know how to get to the North Gate."

Rod was quiet for a moment, processing it all, probably running through the same calculations I had—risk versus reward, necessity versus stupidity, following orders versus staying alive. "I do know how to get there," he finally said, his voice heavy with resignation. "But John... man, I can't believe he opened that door. That's the most insane thing I've ever heard. Opening your door when surrounded by Taliban? What the actual hell was he thinking?"

"You and me both, brother. You and me both."

We drove in silence for a while, both of us scanning the darkened streets for threats, seeing danger in every shadow, every parked vehicle, every figure that moved in the periphery. Rod directed me through a different route, this one wider and better lit, populated with enough late-night traffic that we didn't stand out as obviously as we had on that isolated road. His knowledge of the area came from months of coordinating with IJC, from previous trips I hadn't made.

Within twenty-five minutes, we pulled up to the North Gate of NKAIA. The difference was immediately apparent—this was a proper checkpoint with multiple ANA guards, barriers, light standards, the infrastructure of legitimate security. The guards here were expecting coalition personnel and waved us through after a cursory ID and vehicle check, professional and efficient.

We made our way to the military terminal, a large concrete building near the flight line that hummed with the activity of a 24-hour operation. The place was mostly empty at this late hour, just a few soldiers coming or going on late night flights, plus some cleaning crew working the floor. I approached the desk where a tired-looking staff sergeant sat reviewing flight manifests for the next day, probably counting down the hours until his shift ended.

"We're here to pick up General Williams," I said, already suspecting what he was going to tell me.

The sergeant looked up, his expression shifting to confusion. "General Williams? He left a while ago, sir. Caught a ride with a convoy heading back to HQ ISAF right after the roads opened back up this afternoon. I'm sure he'd be there by now—that was hours ago."

I closed my eyes and let out a long breath, trying to decide whether to laugh or scream. Of course he did. Why would a general wait around in a transient terminal when he had a guaranteed ticket back to HQ with a secure convoy? Any rational person would have done the same thing.

Rod and I looked at each other. He started laughing—not a happy laugh, but the kind that comes when you're so exhausted and frustrated and angry that it's the only response left that doesn't involve violence or crying. The laugh of someone who's reached the end of their rope and discovered it's tied in a noose.

"Let me guess," Rod said, his laughter dying down to bitter chuckling. " Washington doesn't know. He never bothered to check if Williams was still here waiting. Just assumed he'd be sitting around like an obedient puppy."

"Apparently not." I turned to the sergeant. "Thank you, Sergeant. Sorry to bother you."

"No problem, sir. Safe travels back."

We thanked him and headed back to the Cruiser, our boots echoing across the empty terminal. The drive back to HQ ISAF was uneventful, which after the night I'd had felt like a small miracle. We didn't talk much. What was there to say? We'd risked our lives for a mission that had been unnecessary from the start, sent out into hostile territory for no reason, put in danger because nobody bothered to verify the basic facts of the situation.

By the time we got back to HQ ISAF, it was nearly 2200 hours. The base had settled into its nighttime rhythm—fewer people moving around, lights dimmed, the sound of generators more noticeable in the relative quiet.

I found General Washington in his office, still working despite the late hour, hunched over his computer like the evening's events hadn't happened, like we hadn't nearly died on his watch.

"Sir," I said from the doorway, keeping my voice carefully neutral. " General Williams wasn't there."

Washington looked up, irritated—not concerned, not apologetic, just annoyed at the interruption. "What do you mean he wasn't there?"

"He caught a convoy back to HQ ISAF this afternoon, right after the attack when they reopened the roads. He's already here on base, probably been here for hours."

Washington stared at me for a long moment, and I watched something flicker across his face—maybe embarrassment, maybe frustration, maybe just exhaustion. Then he rubbed his eyes and waved his hand dismissively. "Fine. Dismissed."

No apology. No acknowledgment of the danger he'd put us in—twice. No recognition that his poor judgment had nearly gotten us captured or killed. Just "dismissed," like I was bothering him with trivial details rather than reporting on a mission that had gone catastrophically wrong because of his decisions.

I walked back to my barracks in a daze, my legs feeling like they belonged to someone else. The adrenaline was long gone now, replaced by a bone-deep exhaustion that made every step feel like wading through mud. The night air was cool against my face, a relief after hours of stress-sweat inside my body armor.

I thought about my son, about the promise I'd made to come home. About how close I'd come to breaking that promise tonight because of one man's poor judgment and stubborn refusal to admit a mistake. I thought about Marek losing his leg, about the young Army soldiers and advisors killed by the VBIED, the attack at the Supreme Court earlier that afternoon, about all the other names that had accumulated over months of deployment.

I thought about that door opening, those hands reaching in, the moment when survival stopped being certain and became contingent on violence and luck. I thought about the Taliban fighters I'd hit with the Cruiser, about whether I'd seriously injured any of them, and realized I didn't care either way. They'd tried to take us, and we'd survived. That was all that mattered.

That night, lying in my cot in the darkness, I couldn't sleep. Every time I closed my eyes I saw those three Taliban fighters circling our vehicle, their AK-47s in hand, hatred burning in their eyes. I saw the door opening, Washington being dragged out, heard my own voice screaming in my head:

This is how it ends. This is how you die in Afghanistan. Not from an IED or a rocket attack, but from a stupid decision on a dark road.

But it hadn't ended that way. Through luck or skill or divine intervention—probably some combination of all three—we'd made it out. I'd kept my promise to my son, at least for another day. I'd survived another roll of the dice.

As I finally drifted off to sleep sometime after midnight, I made a silent vow to myself: I would never again let someone else's bad judgment put my life at unnecessary risk. Not Washington, not anyone. I still had 140 days left in this kill zone, and I was going to make damn sure I survived every single one of them.

I would be careful. I would be smart. I would question orders that didn't make sense. And if necessary, I would refuse—career consequences be damned—rather than let someone's ego or stupidity get me killed.

No matter what.

Because I had a son waiting for me at home, and that promise I'd made to him was the only thing that truly mattered. Everything else—rank, mission, duty, even survival itself—was secondary to getting home to Jordan.

I would not break that promise. Not for Washington. Not for the mission. Not for anything.

The war would have to find someone else to sacrifice on the altar of bad decisions.

I was done being that person.

Sleep came eventually, and with it, dreams of home.

Chapter 10: Breach

June 25, 2013

The chatter of machine gun fire shattered the pre-dawn stillness at 0430, pulling me from whatever dreamless sleep I'd managed. Before my conscious mind could fully process the sound—that distinctive rapid pop-pop-pop-pop of automatic weapons stitching the darkness—the base-wide speaker system crackled to life with words that sent every deployed military and contractor personnel's heart rate spiking: "DUCK AND COVER. DUCK AND COVER. THIS IS NOT A DRILL."

My hands moved on muscle memory developed over months of deployment—boots yanked on without bothering to tie them, pants pulled up, shirt grabbed, pistol belt cinched. I strapped the 9mm in its hip holster and was out the door before my mind fully caught up with my body, before conscious thought could interfere with survival instinct. The outdoors was already filling with NATO service members in various states of dress, all moving with the same urgent purpose toward their duty stations or defensive positions.

The run to my office felt both infinite and instantaneous, time compressing and expanding in that strange way it does when adrenaline floods your system. My breath came in sharp bursts that had nothing to do with physical exertion and everything to do with controlled panic. All around me, the base was coming alive in controlled chaos—doors slamming, boots pounding concrete, voices shouting commands that were half-swallowed by the hammering of automatic weapons fire that seemed to be coming from everywhere and nowhere at once, echoing off buildings and walls until it was impossible to pinpoint the source.

Then I heard it more clearly—the distinctive crack-whiz-snap of incoming rounds punching against the outer walls, the sound unmistakable to anyone who'd heard it before. Small-arms fire, close enough that I could hear the impacts against concrete T-barriers and the metallic ping of bullets hitting something solid. This wasn't just nearby—this was here, inside our perimeter, breaching the illusion of safety we'd constructed with our walls and wire and weapons. My legs pumped harder, running now instead of jogging.

In the Green Zone. They're shooting inside the most fortified area in all of Afghanistan.

How the hell is this possible?

I hit the office door at a sprint, my hands already reaching for my kit before I'd fully crossed the threshold. Body armor first—the familiar weight settling across my shoulders like an old friend, the Kevlar plates front and back that might stop an AK round if I was lucky and it hit square. Rifle next, the M4 that I'd cleaned a hundred times, that I could disassemble and reassemble blindfolded. My hands moved through the motions automatically: grab the magazine, seat it firmly with the palm of my hand, pull the charging handle back and release it to chamber a round. The metallic chak-chak of a hot weapon.

As I locked the magazine into place and chamber-checked my weapon, Daniel's words from months ago echoed in my head with the clarity of prophecy. We'd sat in the DFAC eating breakfast that tasted like cardboard, and he'd put his hand on my shoulder, his eyes holding that thousand-yard stare that comes from seeing too much.

"It's not if, John," he'd said, his voice carrying the credence of certainty. "It's when. When the base gets hit—and it will—keep your head down. Take cover first, shoot second."

When, not if.

Today was when.

The battle unfolded like a deadly piece of choreographed theater, each act more audacious than the last, revealing a level of sophistication and planning that sent chills down my spine. While I'd been running through the base, heart hammering and mind racing, Taliban fighters had been executing a plan that must have taken months to coordinate—months of surveillance and bribes and forged documents and carefully cultivated insider connections.

They hadn't scaled walls with grappling hooks or cut through concertina wire with bolt cutters. They hadn't tunneled under barriers or blown through gates with explosives. They'd done something far more sophisticated and terrifying: they'd driven through the front gate in Toyota Land Cruisers, displaying official vehicle passes on their windshields— passes identical to the ones sitting in our motor pool, laminated placards that should have been impossible to forge or obtain. They wore Afghan National Army uniforms that appeared genuine down to the unit patches and rank insignia, security badges clipped to their chest rigs, the kind of credentials that would make any checkpoint guard look twice before questioning their presence.

In a war where knowing friend from foe was already the hardest part— where yesterday's ally could be tomorrow's attacker, where trust was a luxury and paranoia was survival—they'd weaponized our own systems of identification and recognition against us. They'd turned our checkpoints and credentials and protocols into access points rather than barriers.

The explosions came in waves that I felt in my chest cavity more than heard—seven, eight, maybe more, it became impossible to count as each blast merged with the next. They erupted right across from my compound, outside the presidential palace where reporters had been gathering for

Afghan President Karzai's morning press conference.[5] Each detonation felt like a hammer blow against my sternum even from my position behind solid concrete.

Suicide bombers, their bodies transformed into weapons of political statement, their deaths designed to create chaos and breach the final perimeter between the Taliban and their targets. What followed the initial explosions was ninety minutes of sustained gunfire—an intensity that made time lose all meaning, where seconds stretched into hours and hours compressed into moments. The remaining fighters had entrenched themselves in buildings overlooking the palace, engaging coalition and Afghan security forces in a close-quarters battle that raged through the most secure blocks of downtown Kabul.

From my concealed position behind the walls of my office building, I tracked the carnage by sound alone, piecing together the battlefield by audio signatures the way a blind person navigates by touch. Each thunderous boom of rocket-propelled grenades slamming next door into the US Embassy Annex, aka Ariana Hotel, sent shockwaves through the concrete beneath my boots, the vibrations traveling up through my legs and spine. That same building where, just seventy-two hours earlier, I'd sat across from Ghost at the Tali-bar, sharing a forbidden drink and classified conversations.

The symphony of violence was unmistakable to anyone who'd spent time in a combat zone: the relentless, rapid-fire chatter of Kalashnikov rifles—that distinctive metallic bark that had become the soundtrack of this godforsaken war—answered by the slower, more deliberate bursts of NATO weapons. M4s and SAWs spitting measured death in controlled three-round bursts, the type of disciplined fire that spoke of professional soldiers fighting for their lives rather than untrained militia spraying bullets in panic.

---

[5] Roggio, Bill. *"Taliban Launch Suicide Assault on Presidential Palace, CIA HQ in Kabul."* *FDD's Long War Journal*, 25 June 2013, *www.longwarjournal.org/archives/2013/06/taliban_launch_suici_1.php*.

The bastards had hit everything that mattered, every symbol of power and authority and the illusion of control we'd spent twelve years trying to maintain. The presidential palace, its ornate walls now bleeding dust and debris, Afghan sovereignty crumbling with each impact. The US Embassy Annex—that fortress of American influence wrapped in concrete and Kevlar, our defiant flag whipping in the breeze even as the compound took fire, as if daring them to try again. ISAF HQ, where coalition flags from twenty-eight nations had flown in manufactured unity, now shrouded in black smoke like a funeral pyre for our pretense of security.

These weren't random targets selected for convenience or opportunity—they were surgical strikes aimed at the very heart of our so-called secure zone, that carefully constructed bubble of safety that encompassed the U.S. Embassy, CIA, HQ ISAF and Afghan Presidential Palace compounds with its layers of razor wire and blast walls and armed guards. The Taliban had proven what we'd always feared but never admitted aloud: all our defenses were just theater, expensive props in a war where the enemy could reach out and touch us whenever they pleased, wherever we thought we were safe.

They'd done the impossible. They'd punched through layers of security that were supposed to be impenetrable, past checkpoints manned by our best troops, past intelligence networks that cost billions, past all our technological superiority and overwhelming firepower. They'd penetrated the one place we'd convinced ourselves was untouchable—the center of our little empire of illusion, now burning under the rising Afghan sun.

When the guns finally fell silent around 0630, the morning air hung heavy with cordite smoke and the acrid smell of burning debris—rubber and plastic and things you didn't want to identify. Eight Taliban corpses littered the battlefield in various states of destruction as the remaining insurgents vanished into the maze of Kabul's back alleys and safe houses, melting away like smoke into the urban landscape where they could blend with civilians and disappear. Three Afghan security guards had paid the ultimate price defending the palace. President Karzai, reportedly inside his office throughout the entire assault, had survived unharmed.

As I stood down and began the process of securing my gear, going through the reverse ritual of de-escalation, my hands were steady but my mind raced. We'd won, if you could call it that—every insurgent killed or fled, no coalition casualties despite ninety minutes of intense combat, the Afghan president safe behind his walls. But they'd proven something far more disturbing: that nowhere was truly secure, that the appearance of safety was just that—an appearance built on wishful thinking and expensive hardware. That fake badges and stolen vehicles and forged credentials could carry death right through our front door, past all our sentries and sensors and supposedly foolproof identification systems.

Daniel had been right all along. It was never a question of if. It was always when, and how bad.

As I walked back across base in the growing light of dawn, stepping past T-walls scarred with bullet impacts—pockmarks where rounds had struck concrete, leaving white divots in the grey surface—I felt something shift inside me. Not fear, exactly, though fear was certainly part of it. Something deeper. A fundamental recalibration of my understanding of this place and my chances of surviving it.

I'd been in Afghanistan for months now, had driven through Kabul's streets countless times, had conducted meetings outside the wire, had witnessed the aftermath of VBIEDs and suicide attacks. But somehow, despite all that evidence to the contrary, some part of me had still believed in the Green Zone. Had still trusted that inside these walls, we were safe. That the concrete and wire and armed guards meant something. That our technology and training and superior firepower created an actual barrier rather than just a psychological one.

This morning had stripped away that illusion like flesh from bone.

The enemy could reach us anywhere, anytime. Not just on Airport Road or outside the wire where we all knew the risks, but here, in the heart of our supposed fortress. They could walk through our gates with the right papers, drive right up to our headquarters in vehicles that looked exactly like ours, wearing uniforms identical to those of our allies.

The war wasn't just "out there" anymore. It had come inside. And if it could happen once, it could happen again.

I reflected on how the war was slowly winding down, one base at a time. Except now I understood that the drawdown didn't make us safer—it made us more vulnerable. Fewer troops meant fewer checkpoints. Fewer patrols meant more opportunities for the enemy to study our patterns. And every base we closed, every position we abandoned, was territory the Taliban could claim the moment we left.

We weren't leaving Afghanistan better than we found it. We were just leaving. And the Taliban knew it, could taste it, were already planning for the day when all these fortified compounds would be theirs again.

Daniel had been right. It wasn't if. It was when.

And eventually, when came for everyone.

Chapter 11: The Cost of Everything

Summer hit Kabul like a furnace door swinging open, the temperature climbing into triple digits before noon and staying there until long after the sun set. The dust turned to powder that hung in the air like talcum, coating everything—your uniform, your skin, your lungs—until you tasted Afghanistan with every breath. And the Taliban, those patient bastards who'd been planning and preparing all winter, stepped up their operations across the country with the regularity of seasonal workers reporting for duty.

They knew we were leaving. Could smell it in the air like predators scenting wounded prey. Every base we handed over, every convoy that pulled out headed north to Uzbekistan or east to the airport, every reduction in troop numbers—it all confirmed what they'd always believed: that time was on their side, that they just had to wait us out, that eventually the Americans would go home and leave Afghanistan to the Afghans, and the Taliban considered themselves the truest Afghans of all.

Fighting season had arrived in earnest, and it brought death with the reliability of the postal service.

I was drowning in work, the responsibilities of being NATO's Chief Basing Officer consuming twelve-hour days that bled into fourteen, then sixteen, seven days a week without respite. There were no weekends in war, no days off, just the endless grinding forward of a machine that demanded constant feeding. My small office in ISAF headquarters had become a war room of spreadsheets and maps, each colored pin representing another compound, another outpost, another piece of ground we'd bled for that we were now signing away with pen strokes and diplomatic language.

I spent my days shuffling through transfer documents written in bureaucratic prose that sanitized violence into administrative procedures, coordinating with Afghan officials who were about to inherit a military infrastructure they couldn't possibly maintain, and trying to make sense of a drawdown that felt more like a controlled panic than a strategy. Every email brought a new crisis. Every phone call revealed another gap in the plan. Every meeting exposed another fracture in the coalition's unity as nations began looking toward their exits.

The work was consuming, all-encompassing, the kind that left no room for thinking about anything else. Which was both a blessing and a curse— I didn't have time to dwell on my near-death experience with Washington, or worry about Jordan back home, or process the accumulating trauma of watching a country collapse in slow motion. But I also didn't have time to breathe, to rest, to do anything except feed the machine that ground men and missions into statistics and reports.

Camp Warehouse was typical of the mess we'd created for ourselves. A sprawling logistics base on Kabul's eastern edge, all concrete barriers and corrugated metal warehouses filled with everything from MREs to spare Humvee parts to medical supplies to ammunition. The compound covered fifteen acres and employed over two hundred Afghan civilians in addition to military personnel. I'd spent weeks overseeing its transfer to the Afghan National Army, walking through endless inventory lists with Afghan officers who nodded politely while I explained supply chain management concepts they'd never implement once we left.

The meetings had been cordial, professional, marked by a diplomatic courtesy that masked deeper frustrations on both sides. The Afghan officers wanted the base—it represented power, resources, legitimacy in a country where all three were constantly contested. But they also resented needing it handed to them like charity, resented the implicit message that they couldn't have built it themselves, couldn't maintain it without our guidance, couldn't be trusted to manage their own security infrastructure.

I understood their pride even as I worried about their capacity. The transfer paperwork was finally signed in late spring, a ceremony attended by various dignitaries who made speeches about partnership and sovereignty and the bright future of Afghan self-determination. The Afghan National Army took full control of the base amid handshakes and photographs, and we added another checkmark to the ever-shrinking list of facilities still under NATO control.

Weeks later, on July 2nd, the Taliban hit it.

I was in my office when the news came through, reviewing proposed timelines for base closures in Helmand Province, trying to coordinate schedules between multiple NATO nations who all had their own priorities and political pressures. The email hit my inbox with the subject line: ATTACK - CAMP WAREHOUSE.

The words on the screen seemed to blur as I read the initial report. They'd used a truck bomb—not surprising, as VBIEDs had become the Taliban's preferred calling card. They'd packed it with homemade explosives—fertilizer, fuel oil, whatever they could scrape together in sufficient quantity to achieve their desired effect. Early estimates suggested 1,200 pounds of explosive material compressed into a vehicle that probably cost them a few thousand dollars to acquire and prepare.

They drove it straight at the main gate at dawn, timing their attack for the shift change when the previous night's guards were tired, and the incoming day shift wasn't fully alert yet. The driver detonated just outside the main entrance, blowing a hole in the reinforced wall big enough to drive three more trucks through—which, according to the report, is exactly what happened next.

Then came the assault team: six or seven fighters dressed in Afghan National Army uniforms—uniforms that might have been stolen, might have been purchased from corrupt supply officers, might have been provided by ANA soldiers who were Taliban sympathizers or just pragmatic men hedging their bets on who'd win this war. The fighters ran through the smoke and debris while the guards were still picking themselves up off the ground, their ears ringing from the blast, their vision obscured by dust and chaos.

The firefight lasted twenty minutes according to the after-action report, though I suspected those twenty minutes felt like twenty hours to the men living through it. The Taliban fighters had shot their way through the entrance with a practiced efficiency that suggested training and experience, killing four Nepalese security guards, two civilian truck drivers, and wounding a dozen more as they advanced toward the main administrative building and the fuel depot—both high-value targets that would maximize damage and disruption.[6]

An Afghan quick reaction force finally cornered them near the fuel depot, and rather than be captured—knowing what awaited Taliban prisoners in Afghan custody—the remaining fighters detonated their suicide vests in a synchronized explosion that collapsed part of the depot structure and sent a fireball climbing into the morning sky.

By the time the smoke cleared and the shooting stopped, there wasn't enough left of the attackers to identify with certainty. Body parts scattered across the compound, mixed with debris and fuel and the charred remains of vehicles. The kind of aftermath that required hazmat teams and body bags and the grim work of trying to reassemble enough pieces to count how many dead.

---

[6] Taliban Hit NATO Supply Compound in Kabul, Kill Six, *ABC13 Houston*, 2 July 2013, https://www.abc13.com/archive/9159144/

The final tally: four Nepalese security guards killed, two delivery truck drivers waiting at the front gate, twelve Afghan Army soldiers seriously injured with wounds ranging from burns to shrapnel trauma to blast-induced brain injuries. Seventeen civilian Afghan employees wounded, most caught in the initial explosion or crossfire. The fuel depot partially destroyed, requiring expensive repairs and replacement of thousands of gallons of contaminated fuel. The main gate compromised, needing complete reconstruction. And the psychological impact—immeasurable but significant—of demonstrating that a base we'd just transferred could be penetrated and attacked within weeks of the handover.

The Taliban had made their point with brutal clarity: We can take anything you give them. We can strike anywhere. Your Afghan allies cannot protect what you leave behind.

I sat at my desk, staring at the casualty report on my screen, and felt the weight of it settle into my chest like a boulder. This was a base I'd personally overseen transferring. I'd walked its perimeter, inspected its defenses, assured the Afghan commanders that the security infrastructure was sound. I'd signed documents certifying that Camp Warehouse was ready for transfer, that the ANA was capable of defending it, that the transition would be smooth.

And only weeks later, the Taliban had walked through the front gate and killed six people.

The question I couldn't shake, the one that kept me up at night staring at the ceiling, was simple and terrifying: How the hell were the Afghans supposed to hold all these bases we were giving them?

We were transferring hundreds of compounds to the Afghan government—combat outposts in Helmand where the Taliban controlled everything outside the wire, patrol bases in Kandahar where every road was an IED waiting to happen, logistics hubs like Warehouse that required constant security and maintenance and operational funding. Each one required guards, utilities, maintenance personnel, ammunition, fuel, food, medical supplies—an endless stream of resources and manpower.

The Afghan security forces were already stretched impossibly thin, spread across a country the size of Texas with a hodgepodge of equipment and inconsistent training and corruption that siphoned off thirty percent of their budget before it reached the troops who actually needed it. Ghost payrolls meant commanders claimed salaries for soldiers who didn't exist. Equipment went missing and showed up in Taliban hands. Officers sold fuel and ammunition on the black market. And now we were handing them an empire of bases they couldn't possibly garrison or maintain.

The money wasn't there—Afghanistan's GDP couldn't support a military the size they'd need. The troops weren't there—recruitment was constant but so was attrition from casualties, desertion, and soldiers simply not showing up for duty. The logistics infrastructure wasn't there—supply chains that barely functioned when Americans were managing them would collapse entirely once we left.

And the Taliban? They were everywhere, patient and predatory, waiting for us to leave so they could take it all back.

I wrote reports about it, detailed assessments with charts and projections showing the math that didn't add up. I briefed generals about it, standing in front of PowerPoint slides that illustrated the widening gap between mission requirements and available resources. Nobody wanted to hear it. The drawdown had its own momentum now, driven by politicians in Washington who'd never set foot in this country and never would, by domestic political pressures that treated Afghanistan as yesterday's war, by a public that had stopped paying attention years ago.

We were leaving whether it made strategic sense or not, whether the Afghans were ready or not, whether it meant abandoning everything we'd spent twelve years building. The only question was how quickly we could process the paperwork and ship the equipment and pretend that we were conducting a responsible withdrawal rather than just cutting our losses and running.

Which brought me to Camp Leatherneck and the absurdity that would define my mid-summer.

The meeting was scheduled for mid-July, a Tuesday that promised to be a scorcher even by Kabul standards—the weather forecast called for 101 degrees, the type of heat that made body armor feel like you were wrapped in a heated blanket while standing in front of an oven. It was being held in the Herat conference room at ISAF headquarters, the same wood-paneled space where we'd held dozens of coordination meetings before. But this wasn't a routine transfer ceremony or a standard coordination session. This was something new, something that made everyone involved uncomfortable because it exposed the cynicism at the heart of our mission: we were going to try to sell the Afghans a piece of infrastructure instead of giving it to them.

The story behind it was classic Washington dysfunction, a bureaucratic comedy that would be hilarious if it wasn't so tragic and if lives didn't hang in the balance.

For years, while the war was still hot and Congress was writing blank checks, they'd approved construction projects across Afghanistan. Bases, training facilities, infrastructure upgrades—billions of dollars in contracts that took forever to negotiate through military and diplomatic channels, longer still to bid out to contractors, and even longer to actually break ground on. By the time contractors actually started building, the policy had shifted. President Obama had announced withdrawal timelines. The American people had made it clear they wanted out. But those construction projects? They were already funded, already started, and stopping them would mean eating the costs anyway—might as well finish and see if we could recoup something.

Camp Leatherneck, the massive Marine base down in Helmand Province that had been the heart of ISAF operations in the south, had been slated for expansion even as we planned to close it. Thirty-four million dollars—thirty-four million American taxpayer dollars—for new facilities.[7]

---

[7] John Sopko, Special Inspector General for Afghanistan Reconstruction, letter to Secretary of Defense Chuck Hagel, July 8, 2013, as reported in "A $34 Million Waste of the Taxpayers' Money in Afghanistan," NPR, July 11, 2013,

State-of-the-art classrooms with climate control and modern learning technology. Shooting ranges with electronic target systems. Barracks with actual plumbing and electricity that worked consistently. Maintenance facilities equipped with tools and equipment worth more than most Afghan villages would see in a lifetime.

The contractors had damn near finished when someone in Congress finally noticed the optics: we were spending millions to build facilities for troops we were planning to withdraw, in a country we were supposedly leaving, at a base we'd already announced would close. The project got halted at ninety-five percent completion, a monument to bureaucratic inertia and political cowardice and the disconnect between Washington timelines and ground reality in Afghanistan.

Now US Forces-Afghanistan and NATO were scrambling to save face and recoup something from the disaster. The solution, handed down from people who'd never spent a day in a combat zone, was elegant in its cynicism: sell the half-finished facility to the Afghan government for five million dollars.

Never mind that we'd spent six times that building it. Never mind that we'd given away hundreds of bases for free over the past two years, entire compounds transferred at no cost because that's what you do when you're withdrawing from a country—you don't charge rent on your way out. Never mind that the Afghans knew this and would obviously see through the charade.

This one—this one we needed to sell, so some congressman could go back to his district and tell voters that we weren't just throwing money away in Afghanistan, that we were being fiscally responsible, that we'd recovered at least a fraction of the investment even though the fraction was laughable and the entire exercise was theater.

---

https://npr.org/sections/thetwo-way/2013/07/11/201195870/a-34-million-waste-of-the-taxpayers-money-in-afghanistan.

The task of making this absurd pitch fell to ISAF Joint Command under Lieutenant General Mark Milley—"Mad Dog Milley" as he was known, though never to his face. IJC would build the presentation and somehow convince Afghan officials to pay for something they knew damn well they'd get for free if they just waited us out. It was an impossible mission assigned to people who were already stretched thin managing actual combat operations.

My job? Simple on paper, nightmare in execution. Coordinate between IJC, Leatherneck, ISAF headquarters, and Afghan officials. Provide oversight of IJC's briefing. Make sure the conference room didn't look like the disaster it usually did. Ensure the video feed actually worked for once instead of cutting out mid-presentation like it did during half our conferences. Verify the translators showed up on time. Confirm everyone arrived with the right rank insignia and the right amount of manufactured enthusiasm for a sale nobody believed in.

It was logistics married to theater, and like every marriage in a war zone, it looked easy from the outside until you were knee-deep in it three days out with everything falling apart simultaneously.

Planning kicked off two weeks before showtime. I made my usual trek to the weekly coordination session with the CIA guys—Ghost, Tom, and sometimes Allen when the topics touched on communications. Our meetings followed a familiar rhythm by now: coffee that tasted like burnt rubber, secure facility that smelled like air conditioning and paranoia, conversations that couldn't be repeated outside the room.

We ground through the routine updates first—which bases were closing when, where Taliban activity was spiking, which local officials were playing ball and which ones were actively screwing us. The Agency didn't just care about security assessments and transfer timelines. They cared about people. Patterns. Who was meeting whom, when, where, and why. The human terrain mattered more to them than the actual terrain, because humans were predictable in ways that geography wasn't.

I waited until we'd exhausted the routine agenda before dropping it into the conversation, keeping my tone casual like it was just another administrative detail rather than what I suspected it might actually be.

"We've got something on the horizon," I said, setting my coffee cup down. "There's a big meeting coming up here at headquarters—Afghan Ministry of Defense, Interior, and Foreign Affairs reps, a couple of ANA generals, ANP brass. We're pitching them on Camp Leatherneck's training facilities."

Ghost's face didn't move—he'd perfected that particular skill over years of covert work where every facial expression could telegraph information you didn't want to share. But something shifted behind his eyes, a subtle focusing of attention like a predator catching scent of prey. "When?"

"July 23rd. 1400 hours."

"Who's on the list?" The question came out flat, professional, but I heard the undercurrent of interest.

I'd been carrying the roster in my cargo pocket for three days, refining it every time IJC changed their mind about who mattered and who didn't. Names, ranks, tribal affiliations—the whole deck of cards from General Milley's IJC Basing office, down through the chain to whatever Afghan decision-makers IJC thought could actually make decisions, though that was always questionable. I slid the paper across the table like I was dealing blackjack in a high-stakes game where the chips were lives and the house always won.

Ghost's eyes moved down the list methodically, that trained intelligence analyst's scan that could extract meaning from patterns most people wouldn't notice. Then they stopped. His finger tapped two names hard enough to leave marks on the paper, his body language shifting from casual interest to focused attention.

"These two. Minister Shah and General Khan. They're locked in?"

"As of yesterday. Why?" I asked, though I already suspected I wouldn't get a full answer. The Agency didn't explain their interest in people—they just collected information and acted on it in ways you'd never see.

He ignored the question with the practiced ease of someone who'd spent a career not answering things. "Shah's portfolio?"

"Infrastructure and development. He's the guy who signs off on base transfers, coordinates with their Ministry of Finance for funding. The money man—or at least the guy who pretends there's money when we all know there isn't."

"Khan?" Ghost was taking notes now in that weatherproof notebook he carried, the kind you could write in during a sandstorm or a monsoon. His handwriting was tight, economical, probably coded in ways I wouldn't understand even if I could read it.

"Corps commander, Helmand Province. He's one of the actual operators who'd use whatever facility we're trying to sell. Boots on the ground type—been fighting in the south for years, probably knows more about Taliban operations than our intelligence analysts do."

Ghost nodded slowly, still staring at the names like they'd just confessed to something. Then his eyes came up to meet mine. "I need specifics. Not just the building—room number, floor plan, access points. Everything."

The prickle started at the base of my skull and worked its way down my spine, that familiar sensation when you realized you were deeper into something than you'd intended to be. "Herat conference room, number 108. First floor, northwest corner. Two entrances, both controlled access with badge checks. No windows. I'll bring you the blueprints next week."

"Security protocols?" He was all business now, the casual friendliness evaporated.

"Standard ISAF. Badge verification at building entrance, mandatory escort for non-ISAF personnel once they're inside. Nobody wanders around alone—too many sensitive areas, too much classified material floating around. Afghan officials will be escorted from entrance to conference room and back out again."

He pulled out that weatherproof notebook and started taking notes with quick, efficient strokes. "Meeting duration?"

"Two hours scheduled. Three if they actually push back on the price, which they will because it's insulting and they know it. We're asking them to pay for something we gave away for free everywhere else."

"They will push back." Ghost closed the notebook with a definitive snap. "These guys aren't stupid. Khan especially—he's been playing both sides since the towers fell."

"Both sides?" The question was out before I could stop it, before I could remind myself that asking for clarification meant learning things I might not want to know.

"Taliban. Us. Whoever looks like they're winning that particular week." He stood up, sliding the notebook into his breast pocket. "Shah's cleaner on paper but he's got connections to Mullah Banadar's network. Nothing we can act on legally, just... connections. The kind you don't sever completely because you might need them later when we're gone and the Taliban are back in charge."

I wanted to push—what exactly did the Agency plan to do with this intel, why did a routine transfer meeting suddenly matter enough for Ghost to take notes like he was planning an operation—but he was already moving me toward the door with that subtle physical guidance that said conversation over.

"Appreciate the heads-up," he said, ushering me to the door. "Keep me in the loop if anything changes on that roster. New names added, people dropping out, timing shifts—I need to know everything."

He escorted me outside the main entrance of the Annex and wished me well. I headed toward my Cruiser and when I looked back, he was gone, disappearing into the Kabul heat like smoke dissipating in the wind.

The night before the big meeting, I stayed late prepping the Herat conference room with Rod and Bill volunteering to help because sitting in the barracks watching pirated DVDs for the hundredth time was marginally more boring than hauling water bottles and arranging furniture. We schlepped cases of bottled water from the DFAC—the type of physical labor that felt good after days of sitting behind a desk pushing pixels around computer screens.

We arranged them on the side table with the obsessive precision of men who knew that officers would notice if anything looked half-assed, and then blame us for ruining the meeting through insufficient attention to beverage placement. Cups stacked in perfect pyramids. Notepads aligned with geometric precision. Pens positioned at exact angles. Details nobody consciously notices unless they're missing, and then suddenly you're the incompetent idiot who ruined international negotiations because the water wasn't cold enough or the pens didn't work.

It was pushing 2100 hours when we finished, the building settling into that hollow, after-hours quiet that made every footstep echo off concrete and tile. Most of the staff had already cycled back to the barracks or were at the DFAC scraping the bottom of whatever mystery meat they'd served for dinner—probably chicken again, always chicken, prepared seventeen different ways that all tasted like disappointment.

We locked up the conference room and headed out the back exit, planning to hit Ciano's—the little pizza joint on base that was run by Nepalese contractors but somehow served a decent approximation of American-style pizza and stayed open late enough to catch the post-work crowd. We were discussing the merits of pepperoni versus sausage when I glanced back toward the building.

Ghost materialized from the shadows near the east wall like he'd been conjured from darkness itself. Not walking casually—moving with purpose, fast and fluid, movement that said mission rather than leisure. He had a small black backpack slung over one shoulder, the kind you could stuff full of surveillance equipment or documents or plastic explosives or whatever covert operators carried these days. He went straight through the same exit we'd just left, moving like he belonged there, like he had every right to be entering a secure facility after-hours.

"I'll catch up with you guys," I said, yanking my phone out and pretending I needed to make a call.

Rod looked back, his friendly face showing concern. "You want us to order for you?"

"Yeah. Pepperoni. Thanks."

They kept walking, their voices fading as they debated whether Ciano's pizza or the pizza back home in their respective countries tasted better—the kind of argument that had no answer but killed time on deployment. I stepped into the shadow of the building's overhang and waited, my heart rate picking up for reasons I didn't want to examine too closely.

Five minutes. Maybe less, though it felt longer standing there in the darkness pretending to be on my phone.

When Ghost came back out, the backpack looked different. Lighter. Flatter. Like he'd unloaded whatever he'd been carrying, left something behind that he didn't need to take with him. He turned west without hesitation, heading back toward the direction of the Embassy Annex compound, moving like a man with a mission accomplished and places to be.

I stood in that shadow for another thirty seconds, mind churning through possibilities. What the hell had he left in there? Audio surveillance devices to record the meeting? Video cameras hidden in air vents or light fixtures? Something to track the Afghans when they sat down, maybe inject malware into their phones if they set them on the table? Or was it something worse—something I didn't want to put words to because once you say it out loud, once you name the thing, you own it and carry responsibility for knowing?

I went back inside, my footsteps echoing in the empty corridor.

The conference room looked exactly the same as when we'd left it. Water bottles in neat rows, labels facing forward. Chairs positioned around the table at regulation intervals. Projector tested and ready, remote control placed exactly where we'd left it. Notepads and pens waiting for tomorrow's attendees. Nothing obviously different, nothing screaming "surveillance equipment here."

I walked the perimeter slowly, methodically, checking behind furniture, under the table, along the window frames. Looking for wires, devices, pinhole cameras, anything that screamed "Agency trickery." I ran my hands under chair arms, checked the undersides of the table, examined the power outlets and light fixtures. Searching for the telltale signs of electronic surveillance—odd wiring, fresh drill marks, equipment that didn't belong.

Nothing.

If Ghost had left something, it was invisible. Professional. An invisibility that meant I'd never find it unless he wanted me to, and maybe not even then. Whatever device or system or technical wizardry he'd installed was beyond my ability to detect, buried in the infrastructure or disguised as ordinary equipment or so small and sophisticated that it might as well have been magic.

I stood there for a long moment in the empty room, staring at chairs that tomorrow would hold Afghan officials making decisions about their nation's future. Then I killed the lights and walked out, closing the door behind me with a metallic click that sounded too loud in the silence.

Tomorrow, Minister Shah and General Khan would sit in those chairs. They'd drink that water, shuffle those papers, examine the glossy presentation materials we'd prepared. They'd make whatever decision they were going to make about whether to buy facilities they knew they could get for free by simply waiting.

And Ghost would be listening. Or watching. Or collecting their phone signals. Or something else I couldn't even imagine because I wasn't trained in the dark arts of intelligence collection and covert operations.

I told myself it didn't matter. The Agency did its thing, and I did mine. Two parallel operations that occasionally intersected but remained fundamentally separate. I was logistics and coordination, they were intelligence and action. Stay in your lane, don't ask questions you don't want answered, and definitely don't look too closely at what the spooks were doing because plausible deniability was a gift they were giving you.

But I didn't sleep well that night, lying in my cot staring at the ceiling while my mind spun through scenarios. What if something went wrong? What if the Afghans discovered the surveillance? What if Minister Shah or General Khan were more important targets than Ghost had let on, and this meeting was actually bait for something bigger?

And the question that bothered me most: How much of my job was actually managing base closures, and how much was providing cover and opportunity for intelligence operations I'd never be briefed on?

July 23rd dawned clear and sunny, the temperature steadily rising. By noon it was uncomfortably hot—a dry heat that sucked moisture from your lungs and made you understand why ancient civilizations had worshipped water. The air conditioning in the conference building was cranking away, fighting a losing battle against Afghan summer and the insulation values of concrete walls built more for blast resistance than thermal efficiency.

I showed up two hours early, triple-checked the room one more time, and coordinated with the translators—two Afghan men who'd worked with ISAF for years and knew the drill better than most Americans. They had that particular wariness that came from being collaborators in a war where collaboration could get you killed after the foreigners left. But they showed up on time, professional, because they needed the money and probably because they'd already calculated their exit strategies for when the Americans finally withdrew completely.

The American and NATO side arrived first, as they always did—punctuality being one of those cultural values we imposed on every interaction. Major General Lieberman led them in, a ramrod-straight German officer with perfect English and the kind of methodical competence that made you believe NATO might actually function as an alliance rather than just a bureaucratic fiction. Behind him came a cluster of colonels and lieutenant colonels from various nations, each representing some aspect of the drawdown—finance, logistics, infrastructure, legal affairs. I was the only major in the room, which made me hyper-aware of every move I made, every word I spoke. In military hierarchies, rank wasn't just about authority—it was about permission to exist in certain spaces.

The Afghans arrived in three separate groups over the next twenty minutes, a fact that spoke volumes about the fractured nature of their government and the tribal loyalties that still trumped national identity. Afghanistan wasn't really a nation in the Western sense—it was a collection of ethnic groups and tribal affiliations and regional powerbrokers who'd been forced into a single political entity by colonizers and wars and international pressure.

Minister Shah came first, flanked by two aides in Western suits that probably cost more than an Afghan soldier made in six months. He was in his fifties, trim and polished, with a carefully groomed appearance that suggested someone who'd learned to navigate both Afghan and American power structures with equal ease. He shook hands, smiled, spoke excellent English with just enough accent to remind you he was Afghan but not so much that it hindered communication. An official who'd mastered the art of telling Westerners what they wanted to hear while quietly pursuing Afghan interests.

General Khan arrived ten minutes later with three other ANA generals, and the energy in the room shifted immediately. These men were different from Shah's polished bureaucrats. Battle-hardened didn't begin to describe it. Khan himself looked like he'd been carved from stone and weathered by decades of war—deep-set eyes that had seen things I couldn't imagine, a thick black beard streaked with gray, scars crisscrossing his face and visible on his hands. He wore an Afghan National Army uniform, but it hung on him like an afterthought, a formality required by his position rather than an identity he'd chosen.

This was a man who'd probably been fighting since the Soviet invasion in the 1980s, a mujahideen warrior who'd learned combat as a teenager and never stopped. He'd probably fought the Soviets, then fought in the civil war, maybe fought for the Taliban before switching sides, now fought against them. A survivor who understood that loyalty in Afghanistan was fluid, that today's ally could be tomorrow's enemy, that the only constant was violence and the need to be on the winning side when the music stopped.

The other generals were cut from the same cloth: older men in their fifties and sixties, faces like leather tanned by sun and hardened by decades of conflict, hands that looked comfortable around rifles but awkward holding pens. These were tribal commanders and warlords who'd been given ranks and uniforms and told to play soldier in a national army, but their loyalties still ran through ethnic lines and regional affiliations rather than the abstract concept of Afghanistan.

The Afghan National Police contingent arrived last—three ANP commanders who looked younger but no less dangerous. These were street fighters who'd clawed their way up through Afghanistan's brutal security hierarchy, men who understood that policing in a war zone meant combat operations wearing different uniforms.

I couldn't help staring at them as they filed in and took their seats at the table. These were not bureaucrats shuffling papers and attending meetings. These were warriors who'd been at war longer than I'd been alive, who carried scars both visible and invisible, who'd survived things that would have broken men from softer places.

And we were about to try to sell them a training facility they didn't need, for a price they wouldn't pay, as part of a drawdown they knew would leave them holding an untenable position against an enemy that had infinite patience.

Major General Lieberman called the meeting to order at precisely 1400 hours with the Germanic precision that characterized everything he did. He spoke in English, pausing every few sentences for the translator who rendered his words into Dari, his tone formal but cordial. He welcomed the Afghan delegation, thanked them for their partnership—words that sounded hollow given what was coming—and outlined the purpose of the meeting: to discuss the transfer and potential purchase of newly constructed training facilities at Camp Leatherneck.

Then he turned it over to IJC, and the absurd theater began.

The video feed kicked in, showing the conference room at Camp Leatherneck two hundred fifty miles southwest in Helmand Province. An Army lieutenant colonel appeared on screen, surrounded by contractors in polo shirts and khaki pants—the unofficial uniform of defense contractors making six figures to rebuild a country that didn't want rebuilding. He launched into his presentation with PowerPoint slides that probably looked impressive to people who'd never actually built anything: square footage calculations, construction timelines, materials specifications, intended purposes broken down by facility type.

Charts and diagrams filled the screen—classrooms with capacity for forty students, shooting ranges with electronic target systems, maintenance bays equipped with hydraulic lifts and specialized tools, barracks with climate control and modern plumbing. All built to American standards, which meant expensive and unsustainable for an Afghan budget that barely covered payroll.

The Afghans listened impassively. Shah took occasional notes with a gold pen that probably cost more than an ANA private made in a month. General Khan sat perfectly still, his scarred hands folded on the table, his eyes never leaving the screen. His face revealed nothing—not interest, not skepticism, not boredom. Just patient observation, the kind that came from decades of listening to foreigners explain things while calculating how to turn the situation to his advantage.

The other Afghan officials showed similar restraint, their expressions carefully neutral. Only their eyes moved, tracking between the screen and Lieberman and each other, conducting silent conversations through glances and micro-expressions that spoke a language I couldn't read.

Fifteen minutes into the presentation, the lieutenant colonel was hitting his stride, walking through detailed specifications of the rifle ranges and their electronic scoring systems—technology that would probably break within a month of Afghan National Army use and never be repaired because the maintenance contracts would evaporate when we left—when he suddenly stopped mid-sentence.

"We've got incoming," he said, his voice changing from presenter to soldier in an instant, professionalism bleeding into tension. In the background through the video feed, I could hear it—the distinct wail of the base siren, that rising and falling alarm that meant indirect fire. Rockets or mortars arcing through the air toward their coordinates. "We need to—"

The screen went black. No fade, no transition—just instant darkness as the video conference system lost connection. Whether they'd cut it deliberately to take cover or whether a hit had taken out their communications, I didn't know. Either way, the presentation had just ended with a reminder that we were discussing theoretical training facilities in a place where people were actively trying to kill each other.

The silence in the conference room was deafening, broken only by the hum of air conditioning and someone shifting uncomfortably in their chair. Major General Lieberman's face remained impassive, but I saw his jaw tighten—the tiny tell of a man maintaining composure through force of will. He turned and looked directly at me, his eyes carrying an unspoken message that my stomach interpreted before my brain could: Fix this.

"Major Spargur," he said, his voice carrying across the room with quiet authority. "Our Chief of Basing will continue the presentation."

My stomach dropped through the floor.

I wasn't prepared to brief. I'd coordinated with IJC, had reviewed the materials, knew the talking points secondhand from planning meetings. But I'd assumed—foolishly, it turned out—that I'd be in the background for this meeting, handling logistics and coordination while the experts presented. Now every eye in the room was on me: NATO generals, Afghan ministers, battle-hardened commanders who'd probably forgotten more about warfare than I'd ever learn.

I stood, forcing my legs to move even though they felt like they belonged to someone else, and walked to the front of the room. The PowerPoint slides were still up on the screen, frozen on a diagram of the facility's layout—a top-down architectural rendering showing how the different buildings connected, where the ranges were positioned, how traffic would flow through the compound. I picked up the remote with hands that wanted to shake and forced them to stay steady through sheer willpower.

"As the lieutenant colonel was explaining," I said, my voice sounding strange in my own ears—too loud, too formal, but at least not trembling— then paused for the translator to render my words into Dari. The ten-second delay gave me time to scan ahead on the slides, figure out what I needed to say next, construct the next sentence while the translator worked.

"The facility comprises six primary structures totaling approximately forty thousand square feet."

Pause. Translate. Use the time to breathe, to think, to prepare the next statement.

"Construction began in late 2011 under a contract funded by Congress as part of the Afghan National Security Forces development program."

Pause. Translate. Keep it factual, keep it moving, don't think about the absurdity of what you're selling.

The rhythm saved me. I could only get two, maybe three sentences out before the interpreter took over, and in those ten or fifteen seconds of translation, I'd scan ahead, figure out what came next, construct my response, and keep my voice steady. I wasn't an expert on construction or training facility design or Afghan security force requirements—but I could read slides and sound competent. I'd been faking confidence for months now, learning to project authority I didn't quite feel, and this was just another performance.

I walked them through the specs, the construction costs sanitized into round numbers that didn't quite capture how much money we'd actually burned. I explained the intended use—training facilities for Afghan National Army forces, classrooms for tactical instruction, ranges for weapons qualification, maintenance bays for equipment sustainment. I laid out why the facility was valuable in terms that sounded reasonable if you didn't think too hard about whether the Afghan military could actually maintain any of it once we left.

And then I got to the price: five million dollars.

The number hung in the air like a fart in church—obvious, embarrassing, impossible to ignore.

I could feel the shift in the room's energy. The Afghan officials' carefully maintained neutral expressions flickered slightly. General Khan's eyes narrowed almost imperceptibly. One of the ANP commanders whispered something to his neighbor behind a raised hand. They all understood what that number represented: an insult disguised as an offer, a slap in the face wrapped in diplomatic language.

We'd given them hundreds of bases for free. Forward operating bases, combat outposts, entire compounds with far more valuable equipment and strategic positioning—all transferred at no cost because that's what you do when you're withdrawing from a country. You don't charge rent on your way out the door. But this one—this particular facility—we needed five million dollars for, and everyone in the room understood exactly why: not because it was more valuable, but because Congress needed political cover.

I pushed through the remainder of the presentation, advancing slides that showed facility layouts and capacity calculations and projected training throughputs—numbers that looked impressive in PowerPoint but meant nothing in the reality of Afghan military operations. The minutes crawled by like hours. My uniform was soaked with sweat, partly from the heat despite the air conditioning, partly from adrenaline and the pressure of presenting to senior officers and foreign officials who could all see through the charade we were performing.

After what felt like an eternity but was probably thirty-five minutes, I reached the final slide. "In conclusion," I said, trying to inject conviction I didn't feel into the words, "this facility represents a significant investment in Afghanistan's security infrastructure. ISAF believes the proposed terms represent fair value for both parties, considering the quality of construction and the strategic importance of maintaining training capabilities in Helmand Province."

Pause. Translate. The interpreter rendered my corporate speak into Dari, probably adding his own internal commentary about how ridiculous this all sounded.

I handed the remote back to Major General Lieberman and returned to my seat, my legs shaking slightly now that the performance was over. My hands trembled as I placed them under the table where nobody could see. I'd done it—gotten through an unprepared presentation without falling apart, maintained professional composure, delivered the message even though the message was garbage.

Lieberman nodded acknowledgment. "Thank you, Major." He turned to the Afghans, his expression carefully neutral. "Gentlemen, the floor is now open for discussion and questions."

The silence stretched for a long moment. Minister Shah broke it first, his voice carrying that smooth diplomatic quality that came from years of practice navigating difficult conversations with foreigners who had money and weapons and too much certainty about how things should work.

"We appreciate the comprehensive briefing," he said through the translator, though his English was probably better than mine and he'd understood every word I'd said. "However, we must ask: why are we being requested to purchase this facility when previous transfers have been conducted at no cost to the Afghan government?"

There it was. The question we'd all been waiting for, asked with surgical precision and impossible to answer honestly.

Lieberman fielded it with practiced diplomacy, explaining the political situation in ways that sanitized the truth into palatable fiction. Congressional scrutiny. Fiscal responsibility. Taxpayer concerns. The need to demonstrate sound stewardship of American resources. It sounded reasonable when he presented it—the careful logic of bureaucratic necessity—but we all knew it was bullshit. The Afghans certainly did.

The real answer was simpler and more cynical: Congress needed political cover. They needed to be able to say they'd recovered some fraction of the money spent in Afghanistan, even if that fraction was laughable and the entire exercise cost more in diplomatic goodwill than it recovered in dollars.

General Khan leaned forward, his voice gravelly and direct, cutting through diplomatic niceties with the bluntness of someone who'd spent too long in combat to bother with euphemisms. When he spoke, the translator rendered his words but couldn't quite capture the edge underneath—the mix of frustration and contempt and weary resignation.

"We do not have five million dollars for facilities we did not request," he said, his scarred hands flat on the table, his eyes locked on Lieberman with unwavering intensity. "If this base is surplus to your needs, you will give it to us when you leave. Or you will destroy it. But we will not pay for buildings that sit in territory we may not control when you are gone."

The last sentence landed like a punch. Territory we may not control when you are gone. An admission of what everyone knew but nobody wanted to say out loud: that the Afghan government's reach extended only as far as NATO firepower allowed, that Helmand Province was contested territory where the Taliban held more ground than official maps acknowledged, that these expensive facilities might end up in enemy hands regardless of who technically owned them on paper.

I spoke up, trying to support Lieberman's position even though I agreed with Khan's assessment. "General, with respect, demolition costs would exceed the sale price. We're offering you functional facilities at a significant discount compared to construction costs—facilities that could enhance ANA training capabilities for years to come."

Khan's eyes shifted to me, and I felt the force of his gaze—the assessment of a man who'd survived decades of war by learning to read people and situations with ruthless accuracy. When he spoke, his voice was quiet but carried across the room with absolute clarity.

"Then destroy it," he said simply. "We will wait."

Three words that ended the negotiation more effectively than an hour of argument could have. We will wait. Because time was on their side, not ours. Because they knew we wouldn't bulldoze thirty-four million dollars' worth of infrastructure out of spite. Because they understood that domestic American politics wouldn't tolerate the optics of destroying facilities we'd just built. Because they could see through every layer of our bullshit to the truth underneath: we were leaving whether they paid or not, and everything we left behind would eventually be theirs by default or Taliban by conquest.

And there it was. They'd called our bluff with the casual ease of poker players who knew exactly what cards we were holding.

Minister Shah smiled faintly, a diplomat's smile that revealed nothing and promised less. "Perhaps we can revisit this matter in the coming months," he said, his voice carrying that particular tone of false optimism diplomats use when they mean 'no' but want to say it politely. "When the timeline for your withdrawal is more clearly defined and we have better visibility into our budget allocations."

Translation: We'll wait you out. Call us when you're actually leaving and need to clear your conscience about leaving stuff behind.

The rest of the meeting was formality—the death dance of a negotiation everyone knew was over but nobody wanted to officially acknowledge had failed. Lieberman tried a few more angles, each one more desperate than the last: payment plans that would spread costs over years, reduced pricing that made the discount even more laughable, joint usage agreements that pretended we'd still be here to share facilities. The Afghans were unmoved, their polite expressions never shifting from diplomatic neutrality even as they quietly refused every offer.

By the time we adjourned at 1630 hours, two and a half hours after we'd started, nothing had been resolved and everyone in the room knew nothing would be. We'd performed our theater, delivered our pitch, gone through the motions required by Congressional mandates and political necessity. And the Afghans had done exactly what any rational actor would do: said 'no thank you' to paying for something they knew they'd get for free by simply being patient.

As the Afghans filed out, Minister Shah shaking hands and making pleasant small talk about coordination and partnership, I caught General Khan looking at me one last time. His expression was unreadable, but I thought I saw something there—not hostility exactly, but a kind of weary recognition. We were both trapped in the machinery of this war's ending, playing roles that didn't quite fit, trying to make sense of decisions made by people thousands of miles away who'd never face the consequences of their choices.

He knew I'd been put in an impossible position, asked to sell something unsellable. And I knew he was being asked to prepare for a future where he'd be expected to fight the Taliban with equipment he couldn't maintain and budgets that didn't exist and allies who were already boarding planes home.

We were all just trying to survive the ending of something that should have ended years ago.

I stayed behind to help clean up the conference room after they left. The water bottles, mostly untouched because Afghans preferred tea and nobody wanted to drink while conducting serious negotiations. The notepads, barely marked with a few polite scribbles. The whole elaborate production felt hollow now, a performance nobody had believed in, theater without an audience.

Somewhere in this room, Ghost's surveillance equipment was still recording or transmitting or collecting data in ways I couldn't detect and shouldn't think about. Either he'd retrieved it before the meeting using some method I hadn't noticed, or it was still here, hidden so well I'd never find it. Some things, I was learning with increasing certainty, you were better off not knowing.

That night, lying in my cot with the base settling into its nighttime rhythm around me, I thought about Camp Warehouse still showing scars from the Taliban attack three weeks earlier. I thought about the hundreds of other bases we were handing over to an Afghan government that couldn't secure them, couldn't maintain them, couldn't afford to keep them running once our money dried up.

I thought about General Khan and those other scarred, hard-eyed warriors who'd be left holding the bag when we finally pulled out completely. Men who'd fought for decades, who'd switched sides and survived purges and navigated the impossible politics of Afghan warfare, now being handed an empire they couldn't possibly defend.

And I thought about that training facility at Leatherneck, sitting empty in the Helmand desert, over thirty million American taxpayer dollars' worth of construction that nobody wanted and nobody would use. In two years, maybe three—probably less if I was being honest—the Taliban would roll back through Helmand Province like a tide coming in, and they'd be using our ranges for their training, sleeping in our climate-controlled barracks, teaching the next generation of fighters in classrooms we'd built.

But at least some congressman could tell his constituents that we'd tried to sell it first. That we'd attempted fiscal responsibility. That we'd done everything possible to recoup taxpayer investment before withdrawing from America's longest war.

That had to count for something.

Didn't it?

Sleep came eventually, but it was fitful and shallow, full of half-formed dreams about empty bases and abandoned equipment and promises made to people who knew they were lies even as we spoke them.

Chapter 12: The Eggers Solution

August 2013

The knock on my door came at 0730—sharp, military sharp, the kind that announced itself as official business rather than friendly visit.

"Major Spargur, the Commander wants to see you. Now."

The words made my stomach tighten involuntarily. I'd been at HQ ISAF for a while now, long enough to understand the hierarchy of summons. When a staff officer wanted to see you, they sent an email. When your direct supervisor needed you, they stopped by your office. But when General Joseph F. Dunford Jr., Commander of all NATO forces in Afghanistan, summoned you personally with the word "now" attached—that meant something significant was about to land on your desk.

I grabbed my notebook and headed outside, weaving across the compound toward the headquarters building. I passed offices where coalition partners from two dozen nations were already deep into their workdays, the low murmur of a dozen languages bleeding through doorways.

Inside, the building pulsed with activity—the type of controlled chaos unique to a military headquarters managing a war that was supposedly winding down but somehow spawned fresh crises each week. Dunford's office sat on the first floor, just to the left after entering the main doors. His aide waved me through without ceremony. The General had no patience for formality when there was work to be done.

Dunford was already standing when I entered, studying a map of Kabul stretched across his desk—a large-scale tactical map marked with colored pins and grease pencil annotations showing security zones, coalition positions, and the ever-shifting boundaries of control. The man was a Marine's Marine, and it showed in everything about him: lean, hard-edged, with salt-and-pepper hair cut high and tight and eyes that carried the thousand-yard quality of someone who'd seen too much but refused to look away. His presence filled the room not through volume or theatrics but through the weight of absolute competence and unwavering focus.

When it came to work, he didn't waste time on pleasantries or small talk—that wasn't his style, and it was one of the things I'd come to respect about him. When Dunford spoke, every word carried purpose.

"Major, we have a problem." He turned from the map, his jaw set in that particular way that suggested he'd already war-gamed several solutions and found them all inadequate. "The Embassy is hemorrhaging personnel. Workers are getting targeted outside the Green Zone—VBIEDs, IEDs, targeted assassinations. The Taliban has been hitting soft targets all summer, and it's only getting worse as we draw down and they sense blood in the water."

I nodded, already knowing where this was heading. Everyone knew it. Kabul had become a hunting ground over the past months, the Taliban demonstrating with increasing boldness that they could reach anyone, anywhere. Embassy workers—translators, administrative staff, contractors, the civilians who kept diplomatic operations running—were particularly vulnerable. They lived in scattered compounds across the city, drove predictable routes, maintained schedules that enemy surveillance could map and exploit. Every month, some of them didn't make it to work.

"Ambassador Cunningham needs housing. Secure housing. Inside the wire." Dunford's finger jabbed at the map, landing on the Green Zone where the Embassy compound sat behind its layers of concrete and armed guards. "The original plan was to consolidate everything at Camp HQ ISAF when we transition to Resolute Support and move operations to Bagram. That was the timeline. But that's 2014, maybe 2015 depending on how negotiations go. State Department needs beds now. Today. Yesterday if possible."

He looked at me directly, and I felt the full force of his attention—the focus that made junior officers check their uniform twice and senior officers prepare more thoroughly. "You're my Chief Basing Officer. I need options. Real options that actually solve this problem. And I need them fast, because every day we wait is another day someone gets killed driving to work."

That's when I understood this wasn't just about real estate logistics or shuffling rooms on a spreadsheet. This was about life and death, about American civilians who'd volunteered to serve their country in a war zone and were now being systematically hunted. Embassy workers—Americans, many with families back home waiting—were getting picked off in traffic, in markets, in their supposedly secure compounds. The Ambassador had dozens of security personnel and administrative staff living in vulnerable positions scattered across Kabul like targets in a shooting gallery, and every morning represented a fresh roll of deadly dice.

I walked out of that office with my guts twisted into knots, the burden of responsibility settling onto my shoulders with physical force. This wasn't theoretical base closure timelines or coordination meetings with Afghan officials who might or might not show up. This was immediate, concrete, life-or-death problem-solving.

For two days, my team and I worked the problem with the focused intensity of people who understood lives hung in the balance. We looked at space projections for every facility under ISAF control, reviewed force flow timelines showing when units would arrive or depart, examined infrastructure capacity calculations for power, water, sewage, everything that determined how many people a compound could actually support. We built spreadsheets and ran scenarios and eliminated options as fast as we generated them.

Camp HQ ISAF was the obvious solution, but Dunford had been clear about that during our meeting: it was a no-go. Despite earlier drawdown estimates that had suggested we'd need less space as forces reduced, the ISAF advisory mission was actually expanding rather than contracting. As we transitioned from combat operations to training and advising the Afghan security forces, we needed more office space, more meeting rooms, more infrastructure to support the growing bureaucracy of nation-building. We'd need every square foot of Camp HQ ISAF to support the Government of the Islamic Republic of Afghanistan ministries we were supposedly empowering. Giving it up early to State Department wasn't an option—it would cripple our own operations.

Then, during a coordination meeting where we were running through our rapidly shrinking list of possibilities, someone on my team mentioned Camp Eggers almost as an afterthought, like they were checking a box rather than suggesting a serious solution.

Camp Eggers.

The name hung in the air for a moment while my brain processed the implications.

Right there in the Green Zone, less than a mile from the main Embassy compound. Heavily fortified with proper blast walls, guard towers, and overlapping fields of fire. Multiple buildings that could house hundreds of personnel. Full infrastructure—power generation, water treatment, dining facilities, medical support, everything needed for sustained operations. And most importantly: it was scheduled for closure at year's end, which meant it would be sitting empty right when the Embassy needed space.

I spent the next twelve hours running numbers with the obsessive focus of someone who'd found a solution that might actually work and couldn't afford to have it fall apart under scrutiny. I confirmed capacity calculations—Camp Eggers could house hundreds of personnel comfortably, more if we doubled up rooms. I checked infrastructure specs—the power plant could handle the load, water treatment was robust, the DFAC could serve hundreds of meals three times a day. I verified defensive positions—the compound had excellent sight lines, reinforced barriers, and access to quick reaction forces.

The facilities were solid, the defensive positions were excellent, proximity to the main Embassy compound was ideal for coordination and logistics. Camp Eggers checked every box on the requirement list.

But there was a catch, and it was a big one that threatened to sink the entire proposal: Camp Eggers sat on multiple parcels of privately-owned Afghan land. Unlike most military bases that occupied government property or land seized through eminent domain decades ago, Eggers was essentially a long-term rental agreement. The Department of Defense was paying millions of dollars every month—millions—for lease rights from various Afghan landlords who'd discovered that renting to Americans was far more profitable than any other use of their property.

If we transferred Eggers to State Department control, they'd inherit those leases and the eye-watering costs that came with them. The monthly bill would shift from DoD's budget to State's, and I had no idea if State could or would absorb that kind of expense. But it was the only viable option that solved the immediate problem of getting Embassy personnel behind secure walls before more of them ended up dead.

When I brought the plan to Dunford two days after our initial meeting, he studied it in silence that stretched long enough to make me nervous. His finger traced the map, calculating distances between the Embassy compound and Camp Eggers, analyzing lines of approach and potential threats, running through the tactical calculus of whether this actually improved security or just moved the problem to a different location. I watched him process every angle with the methodical precision that had earned him his fourth star—security considerations, timing, cost implications, political optics both in Afghanistan and back home.

Finally, he looked up from the map. "This works," he said simply, his voice carrying the finality of a decision made. "Draft a letter to the Ambassador. I'll sign it today."

Relief washed through me, though I kept my expression professionally neutral. Generals didn't need to see their staff officers celebrating when they approved plans—they needed to see confidence that the plan would work.

The letter was direct and professional, vintage Dunford: no bureaucratic padding or hedging language, no diplomatic flourishes or excessive courtesy. Just facts and a clear proposal laid out with military precision. The situation. The requirement. The proposed solution. The implications. The recommendation. Five paragraphs that cut through the noise and presented a decision point that the Ambassador could act on immediately.

When Dunford signed it with his characteristic bold signature—the kind that suggested a man who stood behind his decisions without equivocation—he handed it back to me along with instructions that felt heavier than the paper they came with.

"You deliver this personally. Hand to hand. And you stay there to answer any questions he has about capacity, security, timeline, any of it." He looked at me directly. "The Ambassador needs to trust this will work, and that means you need to convince him it will work. Understood?"

"Yes, sir."

Walking across the Green Zone to the Embassy compound that afternoon, the letter felt like it weighed fifty pounds despite being just a few sheets of paper in a manila folder. The route took me through the familiar checkpoints and serpentine barriers designed to prevent vehicles from building up speed, past the blast walls covered in exhaust stains, through the multiple security perimeters that separated the diplomatic world from the combat zone surrounding it.

At each checkpoint, guards examined my credentials with the kind of thoroughness that suggested recent attacks had made everyone more cautious. The threat level was high enough that even routine movements required deliberate security consciousness. By the time I reached the Ambassador's office suite, I'd been cleared through four separate checkpoints and had my ID verified half a dozen times.

Ambassador James B. Cunningham was a different breed from Dunford, though they shared a similar competence beneath their different exteriors. Where the General was all edges and economy of motion—Marine Corps precision and directness—Cunningham had the smooth, measured presence of a career diplomat. Tall and slender, with thinning brown hair and intelligent eyes, he carried himself with a calm authority that came from decades of navigating international crises where the wrong word could spark conflicts and the right phrase could prevent them.

But that day, when I was shown into his office by his aide, I could see the strain around his eyes that no amount of diplomatic composure could fully mask. He was a man responsible for hundreds of American lives in one of the most dangerous cities on earth, and the pressure was showing in the way he held his shoulders, the lines around his mouth, the carefully controlled tension that suggested someone operating under enormous stress while maintaining the appearance of calm control.

His office reflected his dual nature—part diplomat, part wartime commander. One wall held photographs of him with various international leaders, shaking hands and smiling for cameras that documented decades of diplomatic service. Another wall displayed a map of Afghanistan similar to Dunford's but annotated with different markers—Embassy personnel locations, safe routes updated daily, threat assessments color-coded by danger level.

I handed him the letter without preamble. Military correspondence didn't require extensive introduction—the letterhead and signature said everything that needed saying about who was sending it and how seriously it should be taken. He read it standing up, his expression unreadable as his eyes moved down the page, processing Dunford's assessment and recommendation.

When he finished, he set it down carefully on his desk and looked at me with the kind of direct attention that suggested every detail mattered. "Camp Eggers."

"Yes, sir. It meets all your requirements—security, capacity, timing, proximity to your main compound." I'd rehearsed this briefing during the walk over, organizing the key points in order of importance.

"And the leases?" His voice carried no accusation, just the pragmatic question of a man who needed to understand the full cost before committing to action.

Here was the moment of truth, where the entire proposal could collapse if the numbers were too high or the complexity too great. "Multiple parcels, sir. Private Afghan landowners. DoD currently holds the leases at approximately—" I gave him the monthly figure, watching his face carefully for reaction.

His eyebrows rose slightly—the amount was substantial by any measure—but he didn't flinch or immediately dismiss it. He'd clearly dealt with enough wartime budgeting to understand that security in Kabul came with a price tag that would make civilian accountants weep.

"State would assume those obligations upon transfer?" He was already thinking through the implications, calculating whether his budget could absorb it, what he'd have to give up elsewhere, whether the cost was worth the lives it might save.

"Yes, sir. That's the arrangement in the proposal. DoD transfers the physical facility and infrastructure. State assumes the lease obligations and ongoing maintenance costs." I paused, then added what I thought he needed to hear: "But you'd be consolidating your personnel from dozens of vulnerable locations into one defendable compound. The security improvement alone justifies the expense."

Cunningham walked to his window, hands clasped behind his back, looking out at the sprawl of Kabul beyond the compound walls. Somewhere out there in that maze of streets and compounds, his people were living in unsecured housing, driving through checkpoints controlled by guards of uncertain loyalty, exposed to threats that multiplied as the Taliban sensed American withdrawal approaching. He was doing the same calculation I had—weighing money against lives, short-term costs against long-term security, the certain expense of leases against the uncertain but growing probability of casualties.

He turned back to me after a long moment. "Major, how soon can we begin the coordination process?"

I felt something release in my chest—not quite relief, because there was still enormous work ahead, but the satisfaction of having provided a viable solution to an unsolvable problem. "Immediately, sir. My office is ready to start coordination as soon as you give the word. We can begin facility assessments this week, start the legal review of lease transfers, coordinate with DoD contracting officers to ensure smooth transition. Timeline to occupancy depends on how quickly we can process the paperwork, but I estimate sixty to ninety days if everyone cooperates."

For the first time since I'd entered his office, Cunningham smiled—small and genuine, an expression of relief from someone who'd been carrying an impossible burden and had just been offered a way to put some of it down. "Then let's get to work. I'll have my people contact you today to begin coordination."

He extended his hand across his desk, and I shook it—firm and decisive, the handshake of two professionals who understood they were about to embark on a complex bureaucratic mission that would test both their patience and their competence. "Tell General Dunford I'm more than happy to pick up the tab on those leases. Getting my people behind solid walls is worth every penny and then some."

Over the following weeks, I became the central nexus point for one of the strangest real estate deals in military history—stranger even than the absurd Camp Leatherneck negotiation, because this one actually mattered and actually had to work. I coordinated between ISAF legal teams who specialized in Status of Forces Agreements and military-to-military transfers, DoD contracting officers who'd spent careers managing procurement and property transfers, State Department administrators who understood diplomatic facilities but not military base operations, and Afghan landlords who'd been collecting massive lease payments from the U.S. government for years and wanted assurance the money would keep flowing regardless of which American agency signed the checks.

Every meeting felt like defusing a bomb in slow motion—one wrong word, one bureaucratic snafu, one landlord deciding to renegotiate terms, and the whole thing could collapse. Each stakeholder had their own priorities and concerns. The military lawyers worried about liability and proper transfer protocols. The contracting officers worried about fiscal regulations and audit trails. State Department administrators worried about budgets and congressional notification requirements. The Afghan landlords worried about payment guarantees and what would happen to their leases when Americans eventually withdrew completely.

The magnitude of what Dunford had handed me slowly became clear as I spent sixteen-hour days coordinating meetings and reviewing documents and resolving conflicts between parties who all thought their concerns should take priority. This wasn't just about transferring property or negotiating leases—it was about responsibility for American lives. If I screwed up the timeline, if the coordination failed, if legal issues derailed the transfer, if any of the dozens of moving pieces fell out of alignment, people would remain exposed and vulnerable. People would die because I'd failed to properly manage a bureaucratic process.

It compounded during every conference call, every document review, every negotiation session where Afghan landlords tested our commitment and American lawyers argued over contract language and State Department administrators pushed back on costs. I woke up at 0500 thinking about lease agreements and went to sleep after midnight reviewing transfer protocols. My office became a war room of binders and spreadsheets and timeline charts that tracked every dependency and decision point.

Dunford checked in weekly with questions that were precise and unforgiving, demonstrating he was tracking every detail even while managing an entire war: "Where are we on the lease transfers? Have State's lawyers signed off on the property transfer protocols? What's the latest occupancy timeline? What obstacles remain and how are we overcoming them?"

He wasn't micromanaging—that wasn't his style. He was a commander ensuring his staff officer wasn't drowning under the weight of a critical mission. But the expectation was absolutely clear in every interaction: failure was not an option. American lives hung in the balance, and that meant I needed to make this work regardless of how many bureaucratic obstacles appeared.

Cunningham, for his part, proved to be a steady and effective partner despite our different organizational cultures. State Department and DoD approached problems differently—diplomats favored consensus and lengthy deliberation, military favored decisive action and clear hierarchies—but Cunningham understood that speed mattered here. Despite the diplomatic niceties and careful language, he ran his operation with the same ruthless efficiency as any military command. When bureaucratic obstacles appeared—and they always did, because bureaucracy breeds obstacles like standing water breeds mosquitoes—he cleared them with phone calls that I imagined involved very calm, very firm conversations with very important people in Washington who had the power to make problems disappear.

I never heard those conversations, but I saw the results: legal reviews that normally took weeks got completed in days; budget approvals that should have required congressional notification got expedited through obscure authorities I didn't know existed; Afghan landlords who'd been difficult about lease modification suddenly became cooperative after meetings with Cunningham's staff that I wasn't invited to.

By late fall, the framework was finally in place after months of coordination that had consumed my life. Camp Eggers would transfer to State Department control in a phased operation beginning in November and completing in December. The leases would continue uninterrupted with State assuming all obligations—the multi-million-dollar monthly bill shifting from DoD's books to State's budget without any gap in payment that might give nervous landlords an excuse to renegotiate or withdraw. Embassy personnel would begin moving in before winter, getting behind secure walls before the spring offensive brought renewed violence.

The legal documents filled three massive binders—transfer agreements, lease modifications, property inventories, security protocols, maintenance schedules, utility contracts, everything required to properly hand off a military facility to civilian control while maintaining operational security and meeting both military and diplomatic regulations.

Standing in my cramped office one evening, looking at the mountain of paperwork that represented months of coordination and thousands of hours of work by dozens of people across multiple organizations, I understood something about military service that they don't teach at any academy or in any leadership course: sometimes the most important missions don't involve weapons or combat operations or anything that looks like traditional warfare.

Sometimes you save lives with real estate contracts and lease agreements and bureaucratic persistence. Sometimes the warriors are the staff officers fighting through administrative hell to get people behind walls that can stop bullets. Sometimes survival isn't about shooting straight or moving tactically—it's about coordinating between lawyers and landlords and diplomats until everyone signs the right documents in the right order.

Dunford had given me that mission, and I'd delivered it. Not perfectly—nothing in Afghanistan happened perfectly—but successfully. Embassy personnel would move into Camp Eggers, would live behind blast walls and armed guards, would stop driving vulnerable routes through a city that wanted to kill them. The exact number of lives saved would never be calculated because you can't count the attacks that don't happen, but I knew in my gut that this mattered more than any briefing I'd give or any ceremony I'd attend.

Outside my office window, the Kabul night was punctuated by distant gunfire and the sound of sirens—the ambient soundtrack of a city at war. Somewhere out there beyond the blast walls and checkpoints, Embassy workers were still sleeping in vulnerable compounds, still driving dangerous routes, still hoping they'd make it through another day.

But not for much longer.

Soon, they'd be behind the walls of Camp Eggers. Protected. Secure. Alive.

The war would continue. The Taliban would keep attacking. The drawdown would creep forward with its inexorable bureaucratic momentum. But at least these Americans would be safer, and that was something real in a war where so much felt uncertain and meaningless.

Not on my watch, I thought, gathering the binders and preparing to file them away. Nobody was going to die from preventable exposure on my watch if I could help it.

The mission wasn't glamorous. It wouldn't earn medals or headlines. But it mattered more than anything I'd done in uniform, and I carried that knowledge with quiet satisfaction as I locked my office and headed back to my living quarters for a few hours of sleep before starting again tomorrow.

Sometimes survival looks like combat. Sometimes it looks like paperwork.

Both count. Both matter.

And today, I'd helped ensure more people would survive to see tomorrow.

Chapter 13: Celebratory Fire

Twelve years to the day since the towers fell. The Intel brief that morning had been blunt in a way that made everyone uncomfortable: expect attacks. HQ ISAF was the nerve center of NATO operations in Afghanistan, and the Taliban loved anniversaries with the passion of people who understood symbolic timing mattered as much as tactical success. They'd hit us before on dates that carried meaning—9/11, Afghan Independence Day, the anniversary of Soviet withdrawal—and today was the most symbolic of all, the date that had started this entire war twelve years ago when the towers came down and America decided Afghanistan needed to be invaded.

Every checkpoint was reinforced with additional guards and more thorough screening. Quick reaction forces were on standby with engines already running. We'd increased patrols around the perimeter, tightened security protocols until they bordered on paranoid, and every military and contractor member on base carried their weapon everywhere—to the bathroom, to chow, to meetings, to bed.

The day crawled by in tense silence, every hour passing without incident making people more rather than less nervous. Waiting for an attack you know is coming is worse than the attack itself—the anticipation, the constant state of heightened alert that exhausts you mentally even when nothing physical happens.

Morning turned to afternoon with nothing more threatening than the usual distant sounds of Kabul traffic. Afternoon crept toward evening with the same oppressive nothing, just the usual ambient noise of a city going about its business. No rockets. No suicide bombers. No coordinated assault. Just the normal sounds: distant traffic, the call to prayer echoing across the city from unseen minarets, the periodic rumble of helicopters overhead ferrying people and supplies between bases.

By 1900 hours, I was beginning to think maybe Intel had it wrong for once. Maybe the Taliban had decided to sit this one out, save their resources for a target that mattered more. Maybe they were planning something for tomorrow, or next week, or next month. In Afghanistan, you never really knew—the enemy operated on their own timeline, attacking when it suited them rather than when we expected it.

I hit the DFAC for dinner, my stomach reminding me I'd skipped lunch while managing coordination meetings that had run long. To my pleasant surprise, they were serving lamb—a welcome break from the endless chicken rotations. Rod and Bill's deployments had ended, and I was doing my best to fill in for Bill's lawyer duties while also training Rod's replacement, a Canadian Air Force Captain from Québec. The extra responsibilities left little time for proper meals.

I forced down the lamb with some wilted vegetables that had probably been fresh when they left America weeks ago, and stale bread that required aggressive chewing to break down into something swallowable. I sat alone at a long table while around me soldiers from a dozen different countries ate in exhausted silence. No conversation. No laughter. Just the scrape of forks on trays and the hollow sound of men too tired to care what they were putting in their mouths as long as it counted as calories.

After dinner, I walked back toward my barracks, grateful the day had passed without violence. The September air was cooling down after another scorching day, almost pleasant once the sun set and the temperature dropped twenty degrees. The Afghan twilight was beautiful despite everything—purple and orange bleeding into darkness, the mountains around Kabul turning into black silhouettes against the dying light. It was a natural beauty that made you momentarily forget you were in a war zone, until some sound or smell reminded you exactly where you were.

I was halfway down the street in front of the DFAC, maybe 75 yards from safety, when the world exploded into violence.

Gunfire erupted from every direction at once—not the controlled pop-pop-pop of a firefight, but a massive, sustained roar of automatic weapons fire that seemed to come from the city itself. AK-47s. PKM machine guns. DShK heavy machine guns. The sound was deafening, chaotic, overlapping volleys that seemed to emanate from everywhere and nowhere simultaneously, echoing off buildings and walls until it was impossible to tell where it originated.

This is it.

The thought struck with sudden force, adrenaline flooding my system so fast my hands started shaking. The attack Intel warned us about. After waiting all day, after letting our guard drop just slightly as evening approached, they're hitting us now when we're tired and transitioning to night operations and maybe not quite as sharp as we were twelve hours ago.

The "Duck and Cover" siren started wailing across the base—that long, undulating scream that meant incoming fire, take shelter immediately, this is not a drill. The sound cut through everything else with its electronic urgency, impossible to ignore. Around me on the street, coalition personnel who'd been walking began running toward the nearest hardened structures.

My body moved on autopilot, muscles remembering every drill, every training exercise, every night spent gaming out scenarios. I ran toward my office building. My gear was there—body armor, helmet, tactical gloves, ammunition. I was exposed out here on the street in just my uniform, completely vulnerable to any rounds that might arc over the walls or any direct attack that breached the perimeter.

I heard the crack before I could process what it was. A bullet hit the roof of the building to my right, maybe ten feet above my head and six feet over. The distinctive snap of supersonic rounds cutting through air, that sharp crack-thwip sound that every combat veteran learns to recognize and hate. My brain screamed danger signals that bypassed conscious thought and went straight to the primal survival centers: Move! Run! Get to cover!

I sprinted harder, legs pumping, boots pounding concrete, heart hammering against my ribs. The gunfire was everywhere—a wall of sound that made it impossible to tell where individual shots were coming from. Another round hit somewhere to my left, the thwack of lead on concrete just feet away.

I wasn't thinking anymore in complete thoughts. Just survival imperatives: Get to cover. Grab more ammo. Return fire if needed. The basic calculus of staying alive when people are shooting at you.

I hit the door to my office building and yanked it open. Inside, the fluorescent lights were harsh and surreal after the chaos outside, the normalcy of the hallway jarring against the violence happening meters away. My hands shook as I unlocked my office door—took two tries to get the key in properly because my fine motor control was shot from adrenaline.

I threw on my body armor, strapped on my helmet and chamber-checked my M4 rifle. Locked and loaded. Ready to return fire if anyone came through that door who wasn't friendly.

Then, as suddenly as it had started, the sirens stopped.

Just like that. The wailing cut off mid-note, leaving a ringing silence that was almost worse than the noise had been.

I stood there in my office, breathing hard, adrenaline still flooding my system, trying to make sense of what had just happened. The gunfire outside was still going—sustained, celebratory bursts now that I could actually hear without the siren drowning everything out—but it wasn't directed. Not incoming rounds aimed at targets. Just... firing. Random.

Undisciplined. A type of shooting that suggested celebration rather than combat.

I powered up my computer with fingers that were still trembling slightly, pulled up the Intel feeds, and waited for the reports to come in. Attack reports. Breach notifications. Casualty counts. Something to explain what the hell had just happened.

Nothing. No attack reports. No breach. No casualties.

Then the message came through from IJC intelligence with a subject line that made me stare at the screen in disbelief: **Afghan national soccer team defeats India 2-0 in SAFF Championship final**. Mass celebrations throughout Kabul. All personnel advised that gunfire is celebratory in nature. Maintain security protocols but no hostile action indicated.

I read it twice. Then a third time. The words didn't change.

Not an attack. A celebration.

Afghanistan's national team had just won their first international soccer title—some tournament in Nepal I'd never heard of—and the entire country had lost its collective mind with joy. Hundreds of thousands of people—civilians, Afghan National Police, Afghan National Army, security forces, shop owners, taxi drivers, everyone with access to a firearm—had poured into the streets and started firing every weapon they owned into the air in celebration. AKs, machine guns, pistols, ancient rifles, anything that could shoot. This was victory gunfire, the Afghan way of celebrating, and apparently nobody had thought to warn ISAF that cultural traditions involving automatic weapons fire might be misconstrued as an attack on September 11th.

I sat down heavily in my chair, still wearing full battle rattle, and started laughing. It was either laugh or scream or punch something, and laughing seemed like the option least likely to result in questions from my chain of command. I'd almost been hit—twice, maybe three times judging by how close those rounds came—not by the Taliban, not by an enemy combatant, but by Afghans celebrating a damn soccer game.

This country. This insane, beautiful, deadly country.

Outside, the gunfire continued for another hour, maybe longer. Tracer rounds kept arcing into the sky like fireworks, red and green streaks cutting through the darkness in patterns that would have been beautiful if they weren't fundamentally stupid. Somewhere in Kabul, those bullets would come down—thousands of rounds fired up must come down somewhere. Maybe they'd hit empty ground. Maybe they'd punch through a roof and kill a kid sleeping in his bed, or a mother cooking dinner, or a father walking home from work. But tonight, nobody cared about consequences. Tonight, Afghanistan had won something, and that was worth celebrating with every bullet they had.

What goes up must come down, and in a city as densely packed as Kabul, the odds were good that some of that celebratory gunfire would find human targets on its way back to earth. But that was tomorrow's problem. Tonight was for joy, however dangerous and short-lived.

I took off my gear slowly, the adrenaline crash hitting me hard now that the immediate danger had passed. My hands were still shaking—a fine tremor I couldn't control no matter how hard I tried. My leg muscles twitched involuntarily, still flooded with chemicals designed to help me run or fight. I'd come within feet—inches maybe—of catching a stray round to the head. Not from an enemy deliberately trying to kill me, just from joy. Random chance. Wrong place, wrong time, wrong celebration.

Just another Wednesday in Kabul, where the line between celebration and catastrophe was measured in trajectory angles and bad luck.

Chapter 14: The Weight of Names

We formed up in the courtyard fifteen minutes early for the HQ ISAF memorial unveiling ceremony. Representatives from all twenty-eight NATO countries standing in formation under the Afghan sun that was already climbing toward uncomfortable. Germans next to Poles next to Italians next to Brits next to Americans next to Romanians—a multinational force united in the only ritual that transcended language and cultural differences: remembering the dead.

I stood at parade rest in my OCP uniform outside the Headquarters ISAF entrance, sweat already forming under my collar despite the relatively cool morning temperature. The camouflage pattern felt more appropriate than dress blues somehow—this was a combat zone, after all, even if we were gathered for ceremony rather than combat. Around me, soldiers stood at attention with that particular stillness that comes from muscle memory and repetition—backs straight, eyes forward, hands at sides, the practiced immobility of people who'd done this enough times that the movements were automatic.

Afghan National Security Force officers stood among us in their own distinct uniforms, their bearing matching ours despite the different camouflage patterns. Dignitaries from the Government of the Islamic Republic of Afghanistan had arrived earlier in convoys of armored vehicles, taking their positions near the front where they'd be visible in any photos. This wasn't just another weekly memorial service where we read names and saluted and dispersed. This was the unveiling of something permanent, something that would outlast all of us.

In front of us, the flags of twenty-eight nations hung limp in the still air. No breeze. Just the heaviness of September heat already building, the kind that would be oppressive by noon. Behind the flags stood the new memorial itself, still covered—a black and gold monument waiting to be revealed. Once unveiled, it would stand here permanently, inscribed in English, Pashto, and Dari. A fixed point in a place where everything else seemed temporary.

The ceremony began with music that cut through the morning stillness. The Afghan national anthem played first from speakers set up on either side of the formation, and I watched the ANSF officers stand even straighter, hands over hearts, their faces carrying that complicated mixture of pride and grief that comes from serving a nation that's been at war for longer than most of them have been alive. Then came the NATO hymn, the solemn instrumental that attempted to unite two dozen nations under a single musical banner. The formality of it all—the structure, the protocol, the carefully orchestrated sequence of events—felt like an attempt to impose order on something fundamentally disordered, to make meaning out of chaos through ritual and ceremony.

An Afghan National Army imam stepped forward first, his traditional robes in contrast to the military uniforms surrounding him, followed by the senior NATO chaplain in his formal dress uniform with its chaplain insignia. They would conduct the service of sacrifice together, a joint remembrance that symbolized the partnership we kept proclaiming even as it frayed around the edges. The imam's voice carried across the courtyard as he spoke in Dari, then Pashto, words I couldn't fully understand but whose meaning was clear in his tone and cadence: loss, honor, sacrifice, the theological weight of young men dying far from home.

Then ANA Brigadier General Amin Nasib, chief of religious and cultural affairs, moved to the podium with the measured steps of someone who understood the solemnity of the moment. He was older than most of the soldiers in formation, his beard fully gray, his face lined with a weathering that came from decades of Afghanistan's sun and violence. When he spoke, his voice carried genuine emotion beneath the formality.

"Today is the day we remember those who sacrificed their precious lives for us during Operation Enduring Freedom in Afghanistan," Nasib said, his English accented but precise, each word carefully chosen. "Today we remember those heroes and freedom fighters that paid the ultimate sacrifice to keep us safe and sound."

I shifted my body slightly, careful not to break the position of attention, and felt sweat running down my back. The sun was climbing higher, and we'd be standing here for at least another hour. My mind wandered to the names we'd been reading at these ceremonies' week after week—how many would eventually be inscribed somewhere permanent? How many families were still waiting for that knock on the door, that phone call, that moment when everything changed?

"This memorial," Nasib continued, gesturing toward the shrouded monument behind him with reverence, "will honor the brave men and women who put the needs of others before themselves and their families. Today we come together to recall those who bravely preserved and defended freedom, democracy, and especially our common humanity."

Freedom. Democracy. Common humanity. The words we'd been using for twelve years now, trying to give meaning to sacrifice and purpose to presence. Some days they felt hollow, worn smooth by overuse until they meant nothing. Other days—like today, standing here with Afghan officers who'd chosen to fight alongside us despite the risks, despite knowing they'd be labeled collaborators and marked for death—the words felt almost real, almost worth the cost.

"It is also a day for all of us to hope and pray," Nasib's voice softened slightly, "that one day the true peace we all desire, and that Almighty God desires for us, will indeed come to pass." The hope in his voice sounded genuine despite everything—despite the attacks, despite the Taliban's resurgence, despite the clear trajectory toward our withdrawal and whatever chaos would follow.

He addressed the formation directly now, his eyes scanning the rows of international troops standing at attention in the morning heat. "On behalf of the Afghan armed forces, I would like to thank all contributing nations serving as part of the ISAF team. Your dedication, sacrifice, commitment, and especially the lives of your loved ones and compatriots—those who have paid the ultimate price to see this nation take its place amongst the free nations of the world—will never be forgotten."

I thought about the Army Private First Class, whose name we'd read two weeks ago. About the Marine Corporal. The Army Sergeant. The roster of the dead that kept growing despite assurances the war was winding down. How many more would Nasib have to thank before this was finally over? How many more families would receive folded flags and the hollow promise that their loved one's sacrifice meant something?

"Today I am proud and honored to be with you," Nasib concluded, his voice carrying genuine emotion that the translation couldn't fully capture, "to remember and honor the heroic sacrifices of those service men and women killed or wounded in military action in the effort of securing a free and prosperous Afghanistan."

Next came Bismillah Khan Mohammadi, Afghanistan's Minister of Defense, stepping up to the podium with the bearing of a man who'd spent more time in combat than in government offices. He was broad-shouldered and solid, wearing his uniform with the authority of someone who'd earned his position through action rather than politics. His gray beard was trimmed short in military fashion, and his eyes held that same thousand-yard stare I'd seen in every veteran who'd survived long enough to become old.

He began by offering condolences to the families of fallen ANSF and ISAF service members who had lost their lives fighting terrorism, his voice carrying across the formation with practiced projection. He mentioned the wounded warriors too—wishing them speedy recovery with words that sounded sincere despite being part of the formal script all such ceremonies followed.

Then he pivoted to progress, a move I'd seen hundreds of times in briefings and ceremonies—this balancing act between honoring the dead and justifying the mission that killed them, between acknowledging loss and proclaiming achievement.

"Millions of children go to school," Mohammadi said, his voice gaining strength as he recited accomplishments that were supposed to prove the war had been worth fighting. "Provisions of medical care are expanding across the provinces. More than six thousand kilometers of roads have been paved, connecting communities that were isolated for generations. The Afghan economy is growing at rates that exceed our regional neighbors. These achievements are closely tied to the tireless efforts of ANSF, ISAF, and the international community working together in partnership."

I wanted to believe him, wanted those numbers to mean what they were supposed to mean in the narrative we'd constructed about progress and nation-building. Six thousand kilometers of roads—that sounded impressive until you remembered that many of those roads ran through areas where government control was nominal at best, where the Taliban set up checkpoints at night and collected taxes from travelers, where IEDs turned infrastructure into weapons against the people who built it.

Mohammadi's tone shifted, becoming harder with the edge of defensive pride. "The enemies of Afghanistan have increased their violent activities during the last six months. Their objective was to regain lost territory and bring the achievements of the past eleven years under question." He paused deliberately, letting that sink in, his jaw set with determination. "This is while the ANSF capabilities have steadily increased."

He was making a case, I realized—not just to us but to his own people, to the international community watching from afar, to everyone who needed to believe the Afghan security forces could survive our withdrawal. NATO was gradually pulling out troops according to the timeline everyone knew. The ANSF were assuming the lead for security operations in more districts every month. Everyone was watching to see if they could hold once we were gone, and Mohammadi was projecting confidence even if the reality was more complicated.

"The enemies of Afghanistan thought the ANA by itself would not be strong enough to counter their attacks during this summer's offensive," Mohammadi said, and now there was something like pride cutting through the formal language. "The defeat of the enemy in the past six months has demonstrated the increased capabilities of the Afghan security forces. As a result of the tireless efforts of our forces, the enemies of Afghanistan took all of their objectives to the grave with them."

Bold words. Confident words. I wondered how much was true and how much was necessary rhetoric. Probably some of both—wars were fought with bullets and narratives simultaneously, and the narrative mattered almost as much as the ground. If the Afghan people believed their forces could win, maybe they actually could. If they thought defeat was inevitable, it probably was.

"The key to success of the ANSF is the sincere cooperation of the Afghan public," Mohammadi said, and that at least rang completely true. Without popular support, no amount of training or equipment or NATO assistance would matter. Insurgencies lived or died based on whether the population sheltered them or exposed them.

He ended by returning to the fallen, his voice softening with what sounded like genuine feeling. "I would like to convey my sincere condolences to the families of those who have fallen. I would like to tell their families that their sacrifices will not be forgotten. They have been valuable both to the world and to Afghanistan. May their souls rest in peace."

Then came General Joseph Dunford, commander of ISAF and U.S. Forces-Afghanistan, moving to the podium with that economy of motion that characterized everything he did. He was a Marine through and through, and it showed in his bearing—straight spine, level gaze, the absolute confidence of someone who'd earned every star on his collar through competence rather than politics. When he spoke, his voice carried an authority that didn't need volume to command attention.

"A great leader once said,"[8] Dunford began, his eyes scanning the formation, "Show me the manner in which a nation or a community cares for its dead, and I will measure exactly the sympathies of its people, the respect for their land, and their loyalty to high ideals." He let the quote hang in the air for a moment. "We are here today in the spirit of those words. We are gathered here to pause for a few minutes and reflect on our fallen coalition comrades and Afghan partners. We are here to pay respect to the men and women who demonstrated loyalty to high ideals."

The quote resonated more than I wanted to admit. What did it say about us—about America, about NATO, about this whole enterprise—that we built memorials and read names and held ceremonies? Did it measure our sympathies and loyalty to high ideals? Or was it just the least we could do for people we'd sent to die, the minimum required tribute that let us feel like we'd honored their sacrifice even as we prepared to withdraw and leave behind everything they'd fought for?

"This ceremony is a reminder of the cost of freedom," Dunford continued, his voice carrying authority that transcended the formal language. "The fallen ANSF and ISAF service members made the ultimate sacrifice in battles throughout Afghanistan. All of their lives were tragically cut short before they could fulfill their full potential, before they could return to families waiting at home."

He paused, and in that pause I heard boots shifting on gravel, someone clearing their throat, the distant sound of a helicopter passing overhead.

---

[8] Joseph F. Dunford Jr, "ISAF, ANSF Memorial Unveiled during Ceremony in Kabul" (speech, ISAF Headquarters, Kabul, Afghanistan, September 26, 2013, https://rs.nato.int/newsroom/archive/2013/isaf-ansf-memorial-unveiled-during-ceremony-in-kabul.aspx

The silence held meaning—space for reflection, acknowledgment that words couldn't fully capture what we'd lost.

"I don't believe the focus today should be on how these men and women died," Dunford said, his tone shifting slightly, becoming more personal and less ceremonial. "It's how they lived that's what is important. It is how they lived that makes us remember them with pride rather than just sorrow."

The reframing caught me off-guard, pierced through the emotional armor I'd built up over months of memorial services. I thought about the Army sergeant and specialist from Guam, who'd always had a joke ready and showed everyone pictures of their family back home. I remembered how they lived—the easy smiles, the competence, the plans for when their deployment ended.

Not how they died—not the IED that tore their suburban apart, not the after-action report with its clinical description of the blast pattern and casualty mechanism. How they lived. The people they were before Afghanistan took them.

"These individuals chose to be a part of something bigger than themselves," Dunford said, and I heard the challenge beneath the words. "They chose to accept personal hardship and in many cases great personal risk in order to help bring peace to the people of Afghanistan. That choice, that commitment to service and sacrifice, is what we honor today."

Chose. The word always complicated things. Some had chosen freely, volunteers who'd enlisted because they believed in service or wanted adventure or needed the benefits. Others had been chosen by circumstance—by economics, by lack of options, by the poverty draft that made military service the best or only path forward. But standing here in formation, we all pretended the choice had been clear and the purpose had been pure, because the alternative was acknowledging that some people died for reasons that had more to do with student loans than ideology.

"As we recall the names, the faces, and the stories of the martyrs of Afghanistan and the fallen of the coalition," Dunford continued, his voice taking on the quality of a benediction, "I believe we should ask for God's blessing on those who died and their families that were left behind to carry the burden of loss."

The invocation of God felt appropriate here, in this country where faith was woven into everything—politics, warfare, daily life, the very structure of society. The imam had spoken of God. Nasib had spoken of God. Now Dunford. Different faiths, different languages, different theologies, but the same appeal to something beyond ourselves to make sense of sacrifice and loss and the fundamental injustice of young people dying before their time.

"If we want to honor those who died," Dunford said, his voice taking on an edge now, a challenge directed at every person standing in formation, "each of us should strengthen our personal commitment to the mission in Afghanistan. Each of us must prepare to be a part of a clear and unmistakable message to the enemies of Afghanistan: that no act of violence, no act of terrorism, will diminish our resolve to stand for what is right."

There it was—the transformation of grief into determination, the alchemy that turned loss into motivation. It was what we always did at these ceremonies: turned the dead into fuel for the living, converted sorrow into renewed commitment. It felt necessary and manipulative simultaneously, like most things in war that served multiple purposes at once.

"If we walk away from this ceremony reminded that the cause of freedom requires sacrifice," Dunford concluded with finality, "if we walk away with a renewed sense of commitment to our values, if we walk away reminded of how important it is to defend those values—then I would offer that those who were taken from us prematurely will be able to look down and know that their lives had meaning and purpose."

The speeches ended. Four wreaths were brought forward with ceremonial precision—representing ANSF, ISAF, fallen Afghans, fallen coalition forces—carried by ministry representatives and senior officers who placed them carefully at the base of the covered memorial. The shrouded monument suddenly became the focal point of everything, this hidden thing that would outlast all of us and carry these names into whatever future Afghanistan would have.

A command rang out: "Two minutes of silence."

The courtyard went completely still. Two minutes. One hundred twenty seconds to think about every name, every face, every flag-draped coffin, every family destroyed by a knock on the door. Two minutes that felt both eternal and insufficient—how could two minutes possibly honor thousands of deaths? But we stood there in that silence, each of us alone with our thoughts and memories and the knowledge this war wasn't over yet, that more names would be added before it ended.

A bugler played Taps, the notes cutting through the silence—sharp and clear and impossibly sad, the melody carrying across the courtyard and beyond the walls into Kabul itself. Then a bagpiper began playing a traditional Piper's Lament, the sound lonely and ancient and somehow perfect for this moment. The pipes wailed with a sound that transcended language and culture, expressing grief in notes that every human heart could understand.

The covering came off the memorial with practiced coordination, several soldiers pulling the black cloth away to reveal what had been hidden beneath. Black marble and gold inscription, three languages telling the same story of sacrifice and loss. The monument would stand here outside ISAF Headquarters long after we were gone, after this war ended or didn't end, after Afghanistan became whatever it was going to become. Future soldiers would walk past it. Maybe future Afghan leaders. Maybe someday even tourists if this country ever found peace. They'd read the inscriptions and try to understand what happened here, why so many died, whether it meant something worth the cost.

"Dismissed."

Formation broke with the organized dissolution of military precision. Soldiers peeled off in different directions—some heading toward the DFAC for a late breakfast, others back to offices and duty stations, some just walking alone needing space to process what we'd just witnessed. The Afghan dignitaries moved toward their armored vehicles where security details formed up around them with practiced efficiency. The memorial stood there, newly unveiled, already weathering under the Afghan sun that would age it over years and decades and generations.

I walked back across base alone, taking the long way, needing the extra minutes to think. The question hammered in my head with every step, refusing to be silenced or rationalized away: Was it worth it?

All these names that would be carved into monuments in a dozen countries. All these flags folded into triangles and handed to weeping mothers and stoic fathers and children too young to understand why daddy or mommy wasn't coming home. All these families back home trying to figure out how to keep living after the knock on the door and the chaplain's sympathetic face and the words "We regret to inform you..."

What had we actually accomplished that justified the cost? The Taliban were still here—Mohammadi could talk about defeating them all he wanted, but everyone knew they'd been here for decades and they'd be here for decades more after we left. The Afghan government we'd spent billions propping up was corrupt, fractured, barely functional outside Kabul's city limits. We'd spent trillions of dollars—trillions, a number so large it lost meaning—building an army that might collapse the day we withdrew, training police forces that leaked intelligence to insurgents, propping up a system held together with American money and American blood that would probably bleed out once we stopped transfusing both.

But then—walking past the command building where flags from twenty-eight nations waved in the warm breeze, walking past the motor pool where mechanics were already at work on vehicles that would carry soldiers on missions today—I thought about those six thousand kilometers of roads Mohammadi had mentioned. Even if half were contested, that was still three thousand kilometers of connection. I thought about the millions of children in school, especially the girls who could read now, who could imagine futures their mothers never had. The clinics operating because we maintained enough security for doctors to work without being kidnapped or murdered by Taliban who saw medicine as Western corruption.

Small victories. Incremental progress. Nothing like the clean narrative we'd been sold in 2001 about democracy and liberation and winning hearts and minds. Nothing like the Mission Accomplished banner or the talk of Afghanistan becoming the next South Korea. Just small, grinding, unglamorous progress measured in individual lives slightly less awful than they'd been before.

A school open. A clinic running. A road where civilians could drive without getting blown up every week. Girls learning to read. Women allowed to work. Small freedoms that didn't make headlines but mattered to the people who had them.

Was that enough? Was it worth all those names we'd just honored, all that treasure spent, all those years invested?

I didn't know. Honestly didn't know, and the uncertainty gnawed at me.

I reached my office and sat at my desk—Landry's desk before me, mine now by virtue of fate—and stared at the map of Kabul on my wall. The one I'd marked up with sectors and checkpoints and threat assessments, the visual representation of a city that defied our attempts to control or understand it.

Somewhere out there beyond the blast walls and checkpoints, ordinary Afghans were going about their Saturday, probably not thinking about our memorial ceremony at all. Somewhere else, in compounds we'd never find, Taliban fighters were planning the next attack—the next IED, the next suicide vest, the next ambush that would add more names to next week's casualty report. Somewhere in between those two extremes, people were just trying to survive another day in a country that had been at war for thirty-five years.

An entire generation that had never known peace. Children who grew up thinking gunfire and explosions were just ambient noise, part of life's normal soundtrack.

The memorial would stand outside Headquarters, inscribed in three languages, permanent in a place where nothing else felt permanent. Future generations would walk past it and wonder what we were doing here, whether it was worth it, why we stayed so long and left so little that lasted.

I didn't have answers for them. Barely had answers for myself.

What I had was the mission in front of me—the coordination meetings scheduled for Monday, the lease transfers that needed signatures, the infrastructure reports requiring filing, the small unglamorous bureaucratic work of keeping operations running so we could complete this drawdown with something resembling dignity. Or at least minimize the chaos of our departure.

The monument was unveiled. The names would be inscribed. The families would know their loved ones were remembered, at least here, at least by us, at least in this moment before memory faded and history moved on to other wars and other mistakes.

I owed it to them to believe this war meant something.

Even if most days—especially days like today after memorial ceremonies that laid bare how much we'd lost—I wasn't sure I did.

Even if belief was more about duty than conviction anymore, more about obligation than certainty.

I looked at my computer screen, at the emails already piling up, at the calendar showing another full week of meetings and reports and coordination. Outside my window, another day in Afghanistan was continuing—the same as yesterday, probably the same as tomorrow would be.

The memorial would stand there, permanent, while everything else kept moving forward toward an ending none of us could clearly see.

I rolled up my sleeves and got back to work.

Because that's what survival looked like: putting one foot in front of the other, completing one task after another, making it through one more day.

The war would continue. The drawdown would proceed. And I would do my job until my deployment ended and I could go home to my son.

That was enough.

It had to be.

Chapter 15: Threshold Crossed

October 13, 2013

The October air bit differently than the April cold that had greeted me six months ago. Sharper. Cleaner. A cold that came from snow-capped mountains rather than pre-dawn darkness. I stood outside my living quarters for the last time, watching the sun drop behind the Hindu Kush peaks that ringed Kabul like ancient sentinels. Their summits gleamed white with first snow while the city below still baked during the day and froze at night—the dramatic swings that characterized Afghan autumn.

My replacement, Lieutenant Colonel Martin, had his feet under him now after two weeks of intensive handoff. He was sharp, methodical, organized to the point of obsession—the kind of officer who took notes in three colors and asked questions until he understood not just the what but the why. He'd be fine. Better than fine. The mission was in good hands.

The goodbyes had been completed throughout the day in the way military deployments always ended—matter-of-fact handshakes and brief conversations that acknowledged relationships forged under pressure without getting emotional about them ending. Now, at 1700 hours, I shouldered my assault pack one final time and headed toward the DFAC for a last meal. At 1800, I'd catch my convoy to NKAIA for tomorrow morning's flight out. I was halfway through the serving line when a voice cut through the ambient noise like something from another lifetime.

"John? John Spargur?"

I turned and found myself face-to-face with a ghost from my past. Colonel Edwards stood there in a fresh uniform, looking exactly like what he was—someone who'd just stepped off a transport and was still trying to orient himself to a war zone.

For a moment I just stared, my brain struggling to process the impossibility. Colonel Edwards. My old squadron commander from Malmstrom Air Force Base. A decade ago. Back when I was a young lieutenant responsible for nuclear missiles buried in the frozen plains of Montana, when my life felt stable, predictable, whole in ways I couldn't appreciate until they were gone.

"Sir," I managed, extending my hand. "Colonel Edwards. I can't believe—what are you doing here?"

"Just got in about an hour ago. Assigned to ISAF as the base commander." He gestured at the dining facility with a slight smile. "Saw you in line and thought, 'That can't possibly be Spargur.' But here you are."

"Here I am," I said. "Though not for much longer. I'm flying out tomorrow morning."

His eyes showed genuine interest. "Your rotation's up?"

"Six months. Feels like six years compressed into months."

He studied me for a moment with those sharp commander's eyes that had once evaluated my fitness to control weapons capable of ending the world. Then he gestured toward an empty table in the corner. "How about I join you for dinner?

"I'd appreciate the company, sir" I said, following him to the table."

We settled in across from each other, two Air Force officers separated by rank and a decade of life, thrown together on the last night of one deployment and the first night of another.

"So," Edwards said, cutting into mystery meat with resignation. "Catch me up. What has John Spargur been up to for the last ten years?"

The question hung between us, massive and impossible to answer briefly. I took a long drink of water, buying time to figure out how much truth to tell.

"A lot of assignments," I started. "Left Malmstrom for Cheyenne Mountain—Space Defense Director at the 1st Space Control Squadron. Had a son. Jordan. He's seven now." I paused. "Then Vandenberg—Joint Space Operations Center. That's when everything fell apart."

Edwards waited patiently, giving me space to find the words.

"Jordan developed epilepsy when he was almost three. Severe seizures with no identifiable cause. It's managed now, mostly, but it changed everything. He lives with the constant knowledge that his brain might betray him at any moment."

"I'm sorry," Edwards said quietly, and I heard genuine sympathy.

"My wife had a mental health crisis while we were fostering a little girl, Jayden. One week before signing the adoption papers, social services stepped in." The words came out mechanical because I'd told this story enough times that it had worn smooth. "They told me if I divorced my wife, I could keep our foster daughter. If I stayed married, they'd remove Jayden permanently. So I fought it. Hired lawyers, filed appeals, attended every hearing for three months."

"And?"

"Lost. They took her anyway when she was almost three. Had to watch them carry her out while she screamed for me, not understanding why daddy was letting her go." The memory was vivid—her face, her voice, the look of betrayal. "Filed for divorce after that. Won custody of Jordan. Became a single father."

Edwards set down his fork. "Jesus, John."

"When the orders came for Afghanistan, part of me thought maybe distance would help. Give Jordan stability with my parents. Give me time to process everything." I laughed without humor. "Turns out you can't outrun your life by deploying to a war zone."

"But you made it through," Edwards said. It wasn't a question.

"Made it through. Worked as the NATO Chief Basing Officer—managed base closures and transfers as we draw down. It was consuming. Good to have something that demanded every ounce of focus. Left no room for thinking about anything else."

Edwards nodded slowly. "Sometimes that's exactly what we need."

We talked for another half hour—me filling him in on the realities of Kabul, the coalition dynamics, the frustrations of Afghan bureaucracy. Edwards asked good questions, building his mental map of what he was walking into.

"One more thing," I said, checking my watch. Fifteen minutes. "Trust your Afghan counterparts when it makes sense but verify everything. The corruption runs deep. Not everyone has the same endgame in mind."

"What about threat level? Rockets, suicide bombers?"

"Winter's coming—snow on the mountains means the Taliban scale back. Logistics get harder. You should have a relatively easy few months until late spring." I paused. "But when spring comes, the attacks resume. Stay alert. Don't get complacent. That's when people die."

I checked my watch one last time. "Sir, I need to head out."

Edwards stood, extending his hand. "Thanks for the intel, John. Ground truth from someone who's lived it—that's better than any briefing packet."

"Just passing on what was given to me. Good luck with your tour."

"Same advice I'd give you heading home," Edwards said with a slight smile. "Different battlefield, same principles."

We held the handshake for an extra moment—two officers who'd served together in Montana's frozen missile fields, now meeting in a war zone halfway around the world, each carrying scars the other couldn't see.

"Take care of yourself, John," Edwards said quietly. "And that boy of yours. He needs his father."

"I will, sir."

I shouldered my pack and headed for the exit, leaving Edwards to his dinner—a newly arrived officer beginning the deployment I'd just finished.

Our paths had intersected for one meal, briefly connecting past and present before diverging again.

Outside, the October air had turned genuinely cold. The convoy to NKAIA sat in formation—six up-armored vehicles with engines running, soldiers moving in harsh LED light. The convoy commander checked my name off his manifest.

"Heading home, sir?"

"Heading home."

"Congratulations, sir."

I climbed into the MRAP and found a spot next to a Navy lieutenant and two Army specialists who all wore the same thousand-yard stare—people caught between two worlds, no longer at war but not yet home.

We rolled through Kabul's night streets without incident. No IEDs, no ambushes, just a routine movement through a hostile city that chose not to kill us this particular evening. Thirty minutes later, we pulled into NKAIA.

Hans was waiting outside a Conex near the transient quarters, exactly where he'd said he'd be. Hans, the lead architect for a German firm, a man I'd worked with on countless base redesign projects. We'd become friends over months of meetings—he with his precise engineering mind, me with my military pragmatism.

"Major Spargur," he said warmly. "Your luxury accommodations await."

Inside, the Conex was spartanly furnished but immaculately clean—a cot with actual sheets, a folding desk, even a bottle of water. For someone about to spend their last night in Afghanistan, this qualified as five-star.

"Hans, thank you. Really."

"Is nothing. You fly tomorrow morning, ja?"

"0530. C-130 to Kyrgyzstan, then Germany, and finally home."

"Safe travels, Major. And good luck with your son."

He nodded once—a sharp Germanic nod—and pulled the door closed with a metallic clang.

I stood in the silence, surrounded by quiet for the first time in months. I sat on the cot and pulled out Jordan's photo. I'd promised him I'd make it back before Christmas. Now I was keeping that promise.

Afghanistan hadn't fixed anything. Hadn't healed wounds or brought Jayden back or made Jordan's seizures stop. But it had given me something else—proof that I could still function even when broken. That I could do hard things even when I didn't feel strong.

I'd survived. That was enough.

Exhaustion pulled at me as I lay down, not bothering to remove my boots. My watch alarm jerked me awake at 0445. I gathered my gear in pre-dawn darkness and made my way to the flight line.

The flight line was already busy—ground crews moving, engines warming up, soldiers everywhere heading home or heading in. I joined the formation waiting to board, about forty of us with duffels at our feet. Nobody spoke much. We'd all survived our pieces of this war.

When the Loadmaster called "Major Spargur," I answered "Here," and felt something shift in my chest—relief mixed with anticipation mixed with the strange grief that comes from leaving something difficult but meaningful.

We filed aboard with the awkward waddle of heavily loaded people. The smell hit immediately—JP-8 fuel, hydraulic fluid, old sweat. Months ago, this smell had represented the threshold into war. Now it represented the threshold out.

The engines built to full power, vibration traveling through the aircraft. The rear ramp closed, hydraulics whining as Afghan daylight narrowed and disappeared. This was it. Over six months ago I'd crossed into Afghanistan broken and desperate. Now I was crossing back out still broken but different—tempered by fire, proven capable of surviving what I thought might destroy me.

The engines roared. We accelerated, the nose tilting sharply as we clawed for altitude. We left the ground with a surge, climbing steeply through the moment of maximum vulnerability.

But no alarms sounded. Just steady climb toward safety.

Through the porthole, I caught glimpses of terrain as dawn broke. Kabul spreading across the valley. The mountains rising on all sides, their peaks catching first light, snow gleaming gold and pink. The Hindu Kush stretching away—ancient, indifferent, eternal. The graveyard of empires watching another empire retreat.

The C-130 leveled off. Around me, soldiers shifted into more comfortable positions. The tension of takeoff released like a held breath.

I pulled out Jordan's photograph, studying his face in the red emergency lighting. *I'm coming home, buddy. I kept my promise.*

The thought brought relief but also anxiety. The transition from deployment back to normal life was its own kind of combat. Jordan had aged while I was gone, changed in ways I'd only heard about through brief calls and emails. Would he still need me the same way? Would I still know how to be his father?

An hour passed. Then the pilot's voice came over the intercom, and every head lifted:

"Attention all personnel: we have just crossed out of Afghan airspace. Welcome back to the rest of the world."

The change was immediate. Someone whooped. Someone laughed— genuine laughter, not dark humor but real joy. The tension coiled in everybody suddenly released.

We were out. Safe. Beyond the reach of Taliban rockets and IEDs and all the thousand ways Afghanistan could kill you.

I felt it in my own body—a loosening, an unwinding. My hands unclenched. My jaw relaxed. My breathing deepened.

Out of Afghan airspace. Out of the war.

Epilogue

Three days. Three flights and layovers, then home. Colorado Springs. Jordan waiting—six months older, six months changed, but still my son. Still the reason I'd survived.

Tomorrow was my birthday—October 15th. I'd turn thirty-eight somewhere over the Atlantic, probably eating vending machine food in a German transit facility. I didn't need a party. What I needed was waiting in Colorado Springs.

I thought about that first flight in April, spiraling down into Kabul with my mind full of everything I was running from. I'd arrived broken, barely functional, held together by military discipline and desperation.

I was leaving still broken. The scars hadn't healed—maybe never would. Jordan had been robbed of his mom. Jayden was still gone. My marriage was still destroyed.

But I'd changed. Not healed—healing implied returning to what you were before, and I could never be that person again. Forged, somehow. Tempered. Stronger in the places where I'd fractured because I'd had no choice but to bear the burden I shouldn't have been able to carry.

I'd learned something about resilience they don't teach in leadership courses. It wasn't about being strong enough to avoid breaking—everyone breaks eventually. It was about being stubborn enough to keep functioning after you've shattered. About making it through one more day, then another, then another, until suddenly you've survived months you weren't sure you could survive.

The cargo bay settled into comfortable quiet. Around me, soldiers slept or stared at phones—pictures of family, girlfriends, the people waiting. We were all carrying something. Everyone had their own version of my story— the pieces left behind, the reasons they'd survived.

But we'd made it. We were going home.

The engines droned as the C-130 carried us farther from Afghanistan. Through the porthole, I watched the mountains shrink to shadows. The war receded not with ceremony but with distance—mile by mile, breath by breath—until it no longer pressed against my chest.

I didn't feel triumph. I felt weight. The kind you only notice once you stop carrying it every second.

Afghanistan hadn't healed me. Wars don't heal you. They expose you. They strip away illusions and leave only what remains when comfort and safety are removed. What remained in me was not peace, but resolve.

I'd arrived broken, carrying grief and failure and the quiet fear that maybe I was no longer enough—for my son, for my career, for the life I was supposed to be living. I'd survived not because I was strong, but because quitting wasn't an option. Because responsibility still mattered. Because somewhere far away, a little boy needed his father to come back alive.

Resilience, I learned, isn't heroic. It isn't loud or inspirational. It doesn't look like victory speeches or clean endings. Resilience is showing up when you're exhausted. Making decisions when you're numb. Carrying grief without letting it dictate who you become. It's choosing to live forward even when parts of you are stuck in the past.

I wasn't leaving Afghanistan whole. I was leaving forged—edges sharper, illusions burned away, priorities reduced to their simplest form. Survive. Return. Be present. Protect what still matters.

Ahead of me was another kind of battlefield—parenthood shaped by loss, rebuilding a life that didn't resemble the one I'd imagined, learning how to live without constant vigilance while still carrying its imprint. That fight would be quieter, lonelier, and far longer than any deployment.

But I was ready.

I pulled out Jordan's photograph one last time. For six months, this picture had been a talisman—proof that something good still existed in a world that often felt like it was only breaking and loss.

Now, crossing back out of war, I didn't need it to remind me anymore. The knowledge had settled deeper than images, into something more fundamental. Jordan wasn't just my reason to survive. He was my reason to become someone worth surviving for.

I tucked the photo back into my wallet. I didn't need to keep pulling it out to remember who I was fighting for. That truth was written in my bones now.

The Marine across from me caught my eye and gave a slight nod—the kind of acknowledgment that passes between people who understand what it means to carry someone else's face through a war. He had his own photograph clutched in one hand, his own reason for making it through.

We were all going home to something. The question was whether we could become who those people needed us to be.

I wasn't leaving Afghanistan with answers. I was leaving with better questions.

Not "How do I survive this?" but "What do I build with what survival taught me?"

Not "How do I escape my past?" but "How do I live forward while honoring what I've lost?"

Not "How do I become whole again?" but "How do I become useful while broken?"

Somewhere ahead, three days and eight thousand miles away, Jordan was probably eating breakfast. Maybe telling my parents about a dream. Maybe playing with toys or watching cartoons or doing any of the thousand ordinary things that constitute a seven-year-old's Saturday morning.

He had no idea his father was crossing back over the boundary between war and peace. But he already knew something about loss. About hospitals and absence. About how quickly life can change.

My job now was to make sure he didn't have to keep learning those lessons alone. To stand between him and the world when I could. To absorb what I was able so he could choose when—and how—to carry what was his.

The engines changed pitch slightly—some minor adjustment in our flight path, a correction toward home. I felt it in my chest. Forward motion with purpose. Not running. Not escaping. Just moving toward what waited.

In my wallet, Jordan's photograph rested against my military ID—two images that had traveled a long way together. I didn't need to look at either one anymore.

I knew who I was.

A father, coming home.

About the Author

Maj. Spargur, Kabul, Afghanistan 2013. (Photo by Bill Gibson)

Lieutenant Colonel John Spargur is a retired United States Air Force combat veteran with over two decades of distinguished service to our nation. Born in Cody, Wyoming, he spent his younger years on a small farm in Vermont, attending a country school before pursuing his commission through ROTC at South Dakota State University. Throughout his military career, he worked alongside some of America's most critical defense and intelligence agencies, including the CIA, NSA, NRO, Secret Service, and White House Communications Agency. His unique vantage point from the intersection of military operations, intelligence, and national security gives him rare insight into the realities of service that few can claim.

Now a writer, he brings unflinching honesty to his work, sharing the raw and authentic experiences that shaped his years in uniform. His writing pulls back the curtain on military life—not the sanitized, Hollywood version, but the true stories of sacrifice, complexity, and the weight of decisions made in service to country. For readers seeking genuine perspectives on modern military service and national security operations, his work offers a voice forged in the crucible of real-world experience.

To learn more, visit **https://sites.google.com/view/johnspargur**

Appendix

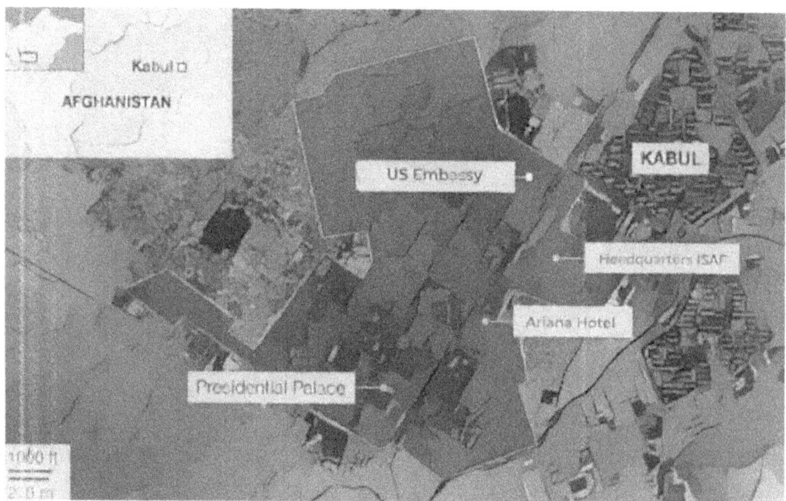

Figure 1: The Green Zone. (Image designed on Canva by John Spargur)